Maths IN THE Real World

Saving and SPENDING

Faye Cowin

NELSON CENGAGE Learning

Australia • Brazil • Japan • Korea • Mexico • Singapore • Spain • United Kingdom • United States

Maths in the Real World - Saving and Spending
1st Edition
Faye Cowin

Cover design: Cheryl Rowe
Text designer: Cheryl Rowe
Typeset: Book Design Ltd www.bookdesign.co.nz
Production controller: Jess Lovell and Siew Han Ong

Any URLs contained in this publication were checked for currency during the production process. Note, however, that the publisher cannot vouch for the ongoing currency of URLs.

First published in 1999 as Putting Maths to Work - Earning Money by New House Publishing

Acknowledgements
Cover image courtesy of Shutterstock.
Images on pages 5,10, 11, 12, 13, 14, 16, 17, 19, 26, 31, 32, 35, 36, 38, 42, 43, 44, 47, 49, 50, 51, 52, 54, 55, 58, 59, 60, 61, 65 courtesy of Shutterstock.

National Library of New Zealand Cataloguing-in-Publication Data
Cowin, Faye.
Saving and spending / Faye Cowin.
(Maths in the real world)
Previously published in the series: Putting mathematics to work. Auckland, N.Z. : New House Publishers, 1997.
ISBN 978-0-17-021710-1
1. Arithmetic—Problems, exercises, etc.
2. Finance, Personal—Mathematics—Problems, exercises, etc.
I. Title. II. Series: Cowin, Faye. Maths in the real world.
513.076—dc 22

Cengage Learning Australia
Level 7, 80 Dorcas Street
South Melbourne, Victoria Australia 3205

Cengage Learning New Zealand
Unit 4B Rosedale Office Park
331 Rosedale Road, Albany, North Shore 0632, NZ

For learning solutions, visit **cengage.com.au**

Printed in China by China Translation & Printing Services.
1 2 3 4 5 6 7 16 15 14 13 12

Maths IN THE Real World

Saving and SPENDING

Faye Cowin

Australia • Brazil • Japan • Korea • Mexico • Singapore • Spain • United Kingdom • United States

Contents

Introduction

Maths in the Real World is a series of theme based books identifying the necessary mathematical skills and knowledge needed in particular areas for the future. The books cover key aspects of the *New Zealand Maths and Statistics for New Zealand curriculum.* The skill requirements of the NCEA Level One achievement standards and remaining unit standards are covered, enabling students to be examined in these NCEA qualifications. They also address the requirements of the Numeracy Project.

Maths in the Real World is aimed at those students who wish to pursue a non-academic career but for whom mathematics is an essential component of their trade training or life in the future. The series is an alternative course in mathematics for 15 to 17-year-olds in schools or training centres. Together, the books offer students:

- Practice with basic mathematics and calculation skills, so essential in the transition from school to the next stage in their development – flatting, travelling, working, buying a car.
- Opportunities to apply mathematics in practical everyday situations – making budgets, shopping, earning money, paying bills, planning a trip, owning a car.
- An awareness of their individual rights and responsibilities and an introduction to the range of community facilities available to them.

Essential skills covered in this book

- Whole number and all operations
- Rounding of decimals
- Form filling
- Problem solving
- Decimals and all operations
- Percentages

ISBN: 9780170217101

New Zealand curriculum level 5: Numbers and algebra

Number strategies and knowledge

- Reason with linear proportions.
- Use prime numbers, common factors and multiples and powers (including square roots).
- Understand operations on fractions, decimals, percentages and integers.
- Use rates and ratios.
- Know commonly used fraction, decimal and percentage conversions.
- Know and apply standard form, significant figures, rounding and decimal place value.

Measurement

- Select and use appropriate metric units for length, area, volume and capacity, weight (mass), temperature, angle and time, with awareness that measurements are approximate.
- Convert between metric units using decimals.
- Deduce and use formulae to find the perimeters and areas of polygons and the volumes of prisms.
- Find the perimeters and areas of circles and composite shapes and the volumes of prisms including cylinders.

Statistics

Statistical investigation

- Plan and conduct surveys and experiments using the statistical enquiry cycle:
  - determining appropriate variables and measures
  - considering sources of variation
  - gathering and cleaning data
  - using multiple displays, and recategorising data to find patterns, variations, relationships and trends in multivariate data sets.
- Comparing sample distributions visually, using measures of centre, spread and proportion.
- Presenting a report of findings.

Statistical literacy

- Evaluate statistical investigations or probability activities undertaken by others, including:
  - data collection
  - methods
  - choice of measures
  - validity of findings.

ISBN: 9780170217101

Literacy and Numeracy Standards for NCEA Level One

NCEA Level One Numeracy Standard	Number	Credits	New Zealand Curriculum
Apply numeric reasoning when solving problems	91026	4	NA6.1: Apply direct and inverse relationships with linear proportions NA6.2: Extend powers to include integers and fractions NA6.3: Apply everyday compounding rates
Solve measurement problems	91030	3	GM6.2: Apply the relationships between units in the metric system, including the units for measuring different attributes and derived measures GM6.3: Calculate volumes, including prisms, pyramids, cones and spheres, using formulae
Solve measurement problems involving right angled triangles	91032	3	GM6.1: Measure at a level of precision appropriate to the task GM6.5: Recognise when shapes are similar and use proportional reasoning to find an unknown length GM6.6: Use trigonometric ratios and Pythagoras' theorem in two and three dimensions
Apply transformation geometry	91034	2	GM6.8: Compare and apply single and multiple transformations GM6.9: Analyse symmetrical patterns by the transformations used to create them.
Use the statistical enquiry cycle to investigate bivariate numerical data	91036	3	S6.1: Plan and conduct investigations using the statistical enquiry cycle
Solve problems which require calculation with whole numbers	8489	2	Solve problems which require calculation with whole numbers (Expires Dec 2012)
Solve problems using calculations with numbers expressed in different forms	8490	2	Solve problems using calculations with numbers expressed in different forms (Expires Dec 2012)
Read and interpret information presented in tables and graphs	8491	2	Read and interpret information presented in tables and graphs (Expires Dec 2012)
Use measurement devices to measure quantities	8492	3	Use measurement devices to measure quantities (Expires Dec 2012)
Find relationships between measurements	12319	2	Find relationships between measurements (Expires Dec 2012)
Make estimates of measurements with common units	20662	2	Make estimates of measurements with common units (Expires Dec 2012)
Use numeracy strategies to solve problems involving whole numbers	23738	2	Use numeracy strategies to solve problems involving whole numbers (Expires Dec 2012)
Use numeracy strategies to solve number problems involving decimals, percentages and fractions	23739	2	Use numeracy strategies to solve number problems involving decimals, percentages and fractions (Expires Dec 2012)

ISBN: 9780170217101

Saving and spending

People live on the income they receive. Some people make goods and sell them, such as manufacturers, while others sell their expertise, such as doctors. Some people get an income without having to physically work for it, for example from investments and benefits.

What is Internet banking?

What is direct debiting?

What age do I have to be to own a credit card?

What is phone banking?

What is an automatic payment?

Can anyone borrow money from a bank?

What does it mean to reconcile your account?

Do I need a cheque book?

What is an overdraft?

ISBN: 9780170217101

Banking

There are three main types of bank:

- **Savings banks**, which include KiwiBank, trustee savings banks and building societies. These are more suited to an individual customer's needs.
- **Trading banks**, which include ANZ, BNZ, National Bank and Westpac. These may be more suited to business needs, but they also have many banking options available for individuals.
- The **Reserve Bank**, which is the government-owned bank in Wellington.

Savings and trading banks are the two we will be dealing with, so let's have a closer look at what they offer. Both offer similar services, however, you should choose the one that best suits to your needs.

Banking services

Most major banks today have special accounts that are suitable for young people up to the age of 19 years. These accounts vary in their emphasis on saving, and others charge for frequent transactions, so when selecting an account you need to consider all the features.

- How much interest will you earn?
- What fees will you need to pay?
- Do you get charged for using an ATM (Automated Teller Machine) that belongs to a different bank?

ISBN: 9780170217101

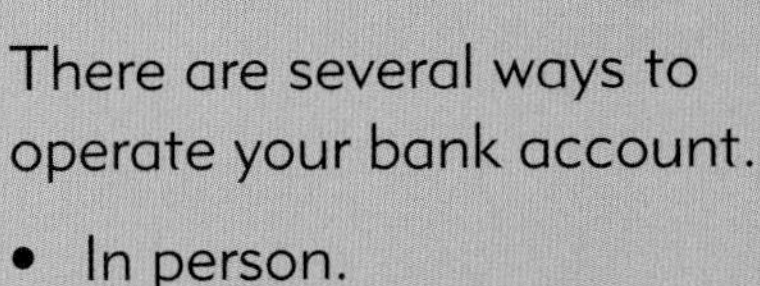

There are several ways to operate your bank account.

- In person.
- Using the Internet.
- Over the phone.

Internet banking and/or phone banking are types of electronic banking. It allows you to manage your account(s) 24 hours a day, 7 days a week, anywhere in New Zealand or overseas. It is easy, convenient and simple to use. You can pay accounts to another person or business (provided you have their account details), transfer money between your accounts and it is secure. Many businesses now prefer to be paid electronically. Your local bank will help you set up Internet and/or phone banking, providing you with a password and other necessary details.

1 Accounts

Savings and trading banks have a variety of different account types available, to meet the needs of different members of the community. The following information is a general outline of some of the available types.

Everyday accounts

These make day-to-day banking easy to manage. Everyday accounts usually charge low or no fees and, provided you only use ATMs of the issuing bank, there are no fees for ATM transactions. You have access to and control of your account 24 hours a day. An EFTPOS card can be issued with these types of accounts. Most banks have special student savings accounts, which enable you to manage your finances while you are a still studying, and several also offer interest-free overdrafts.

Savings accounts

These special savings accounts reward regular deposits. Some accounts have other features such as no transaction fees, however, they may charge a fee for making a withdrawal. You have access to the account at anytime via the Internet or by phone.

ISBN: 9780170217101

Cheque accounts

Many businesses and companies still use cheques, however, electronic transactions are becoming more popular. Cheque accounts are usually fee and interest-free.

Term deposit accounts

This is a special type of savings account. The term of an investment deposit can vary from one month to many years, and all reward customers by paying interest according to the type and amount of the investment. The minimum amount you can put on term deposit is usually at least $5000.

2 Personal loans

Provided you are able to justify the need for the loan – and of course repay it – a bank will allow you to borrow their money. You might need a personal loan for any number of reasons, such as for travel or to make a special purchase, to get your car fixed or to fund home improvements.

Personal loans are available at a range of interest rates. They can be repaid weekly, fortnightly or monthly. Loan approval usually costs around $200. Loans are more commonly taken for amounts greater than $3000.

3 Credit and bank cards

You need to be at least 18 years of age to own a credit card, such as a Visa or Mastercard. They can be cost effective when used everyday and are ideal for managing your finances provided the balance is paid in full each month. Most will charge an annual membership fee. The current interest rate for credit cards is 19.95% pa (per annum or per year), and 22.2% pa for cash advances.

EFTPOS cards give you access to everyday banking, savings or credit card accounts 24 hours a day, 7 days a week. EFTPOS is used at ATMs and EFTPOS outlets. You can withdraw up to $2000 from an ATM using this card, and up to $3000 for an EFTPOS transaction.

ISBN: 9780170217101

4 Insurance

Banks can offer you life insurance as well as lifestyle, travel, contents, car, property and personal insurance policies.

5 Travel and foreign exchange

Banks provide services for overseas travellers such as supplying foreign currency cash and travel cards, making international payments and arranging travel insurance, telegraphic transfers and international bank drafts.

6 Home loans

A broad variety of home loan options are available to meet the needs of their customers.

7 Investments

Banks provide financial advice and planning for investments, shares, KiwiSaver and Bonus Bonds.

ISBN: 9780170217101

Operating bank accounts

When you wish to open a bank account you will need to visit the chosen bank to discuss your requirements. Once you have decided on the type of account(s), the bank will give you the necessary information on how to operate the account(s) and most importantly, your chosen security password and number. It is essential these security details are memorised and kept in a secure place – not in your wallet with your bank or credit card!

Operating a savings account

Most people will have at least one savings account. Savings accounts offer different benefits but their operation is basically the same. Most accounts can be operated electronically or by telephone which is a convenient and safe method. However, you still need to know how to complete a deposit slip and be able to read and reconcile a bank statement (see page 21). A cheque is usually deposited into an account.

On the next page there is an example of the front and back of a deposit slip. The front outlines details of the account holder and a summary of the deposits, which can be cash or cheque(s). The reverse side is for details of the drawer(s) of the cheque(s) which are being deposited. This total must correspond with the cheque total on the front side. Anyone can deposit money into another person's account, provided they have the necessary details of the account.

ISBN: 9780170217101

You need to take note of the following points on a deposit slip below.

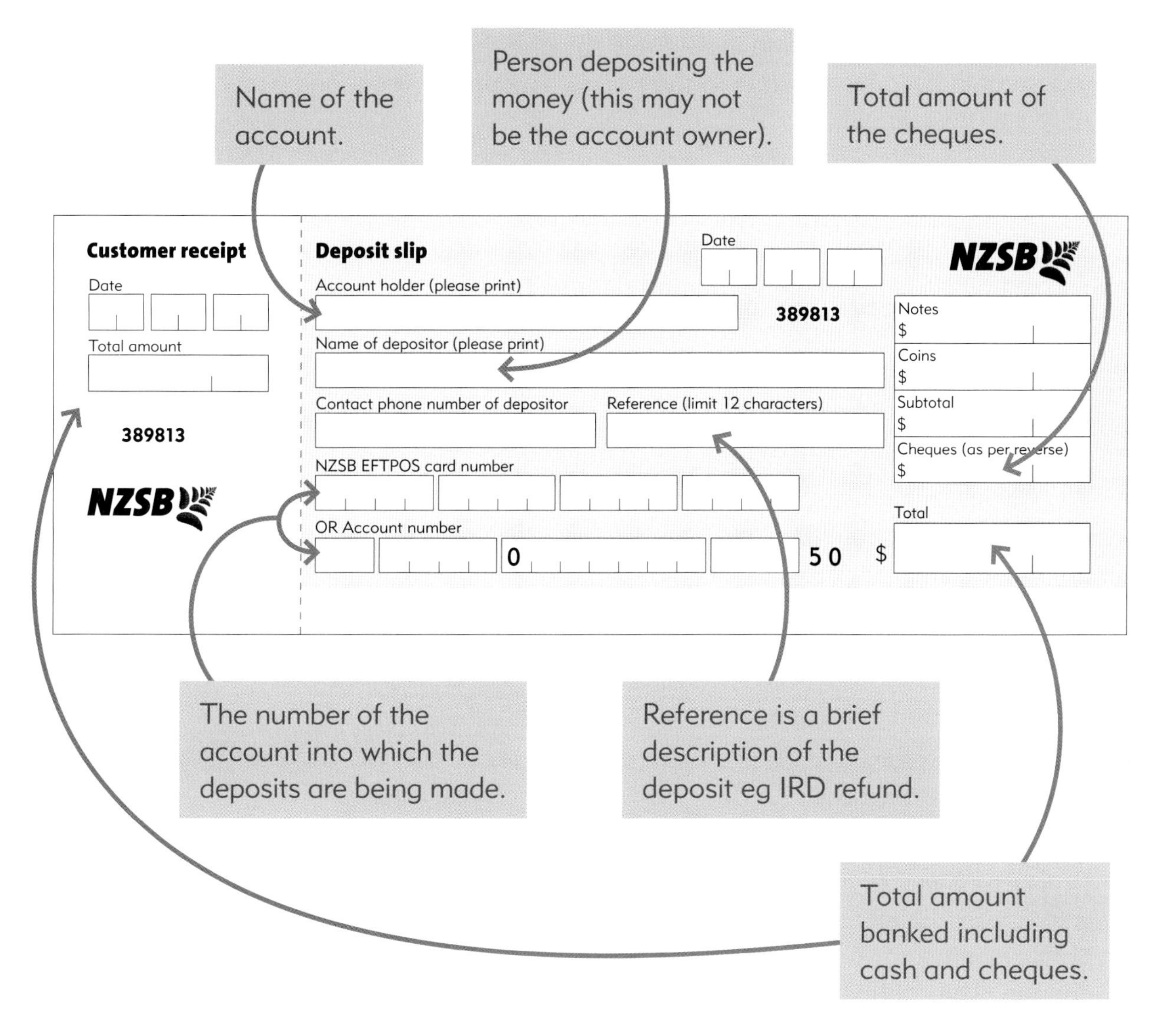

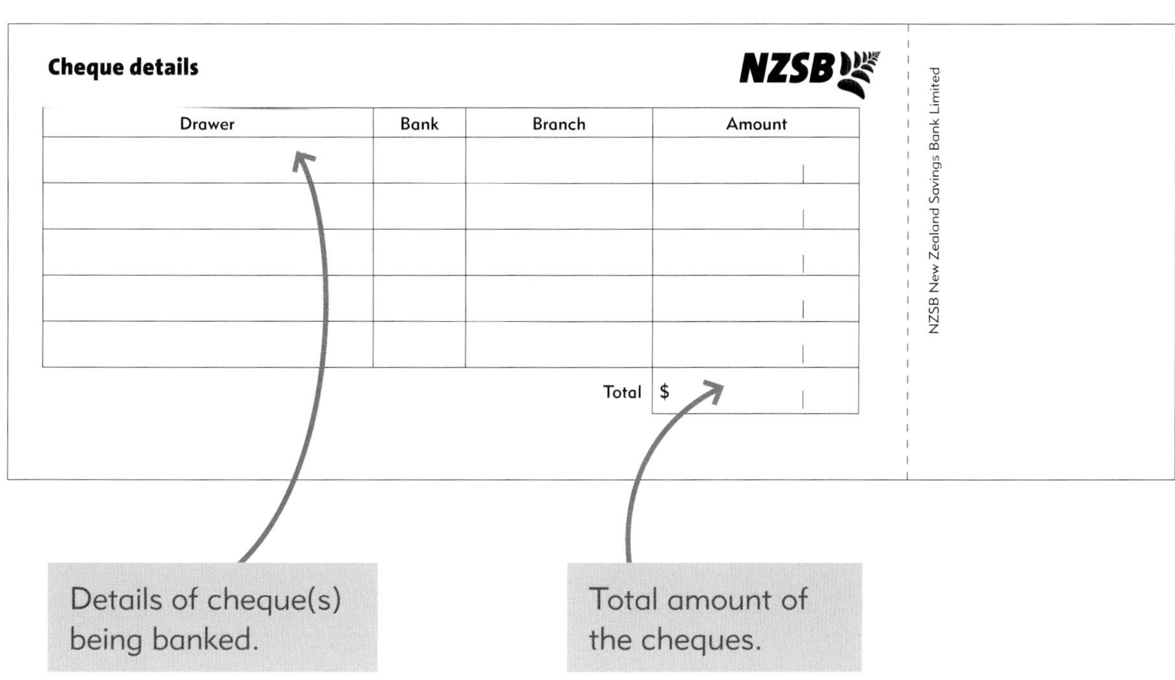

Details of cheque(s) being banked.

Total amount of the cheques.

ISBN: 9780170217101

Operating a cheque account

Cheque accounts are considered to be a safe, portable and convenient method of banking. Many businesses and individuals still operate cheque accounts, however, cheques could become obsolete in the future. To operate a cheque account you must be at least 18 years old.

The cheque account will give you a number of benefits:

- It allows you to use Internet/phone banking and all its benefits.
- You will be issued with a **cheque and deposit book**.
- Direct crediting and automatic payment facilities will be available to you.
- Transactions are recorded on a **bank statement** that the bank posts to you, or you can view them online.
- There is (usually) no interest paid on a cheque account.
- You will be entitled to a **credit card**.
- Provided certain conditions are met you may be entitled to **overdraft** facilities.
- You will be able to have an EFTPOS card.

Whenever you write out a cheque, it is essential that you remember these points:

- It must be dated (cheque is valid for six months).
- The written words must be the same as the digits. If not, the bank will pay out on the smaller amount.
- It must be signed.
- One secure way to make the money payable only to a specific person or business is to 'cross' the cheque. This means to write the words **not transferable** within twin parallel lines that you mark, running diagonally in the top left hand corner of the cheque. The cheque must then be paid into that particular account.
- If you make any mistake on the cheque, it must be initialled by you.
- If you wish to cash a cheque it is wise to write it out for the amount at a bank. This type of cheque is called an **open cheque**.
- On the cheque below, discuss who is the payee, drawee or drawer.

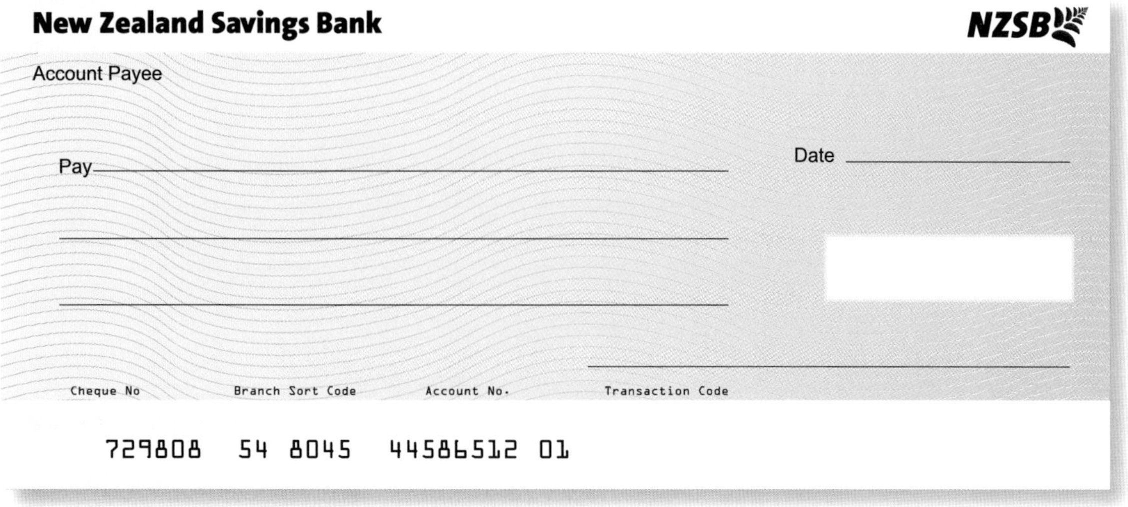

New Zealand Savings Bank NZSB

Account Payee

Pay ____________________ Date ____________

Cheque No Branch Sort Code Account No. Transaction Code

729808 54 8045 44586512 01

ISBN: 9780170217101

This is a deposit slip for a cheque account. You will notice it is the same as the savings account deposit slip on page 15, so the same important points apply.

Customer receipt

Date

Total amount

389813

NZSB

Deposit slip

Date

NZSB

Account holder (please print)

389813

Name of depositor (please print)

Contact phone number of depositor

Reference (limit 12 characters)

NZSB EFTPOS card number

OR Account number

0 | 50

Notes	$
Coins	$
Subtotal	$
Cheques (as per reverse)	$
Total	$

||ı50

Cheque details

NZSB

Drawer	Bank	Branch	Amount
		Total	$

NZSB New Zealand Savings Bank Limited

Exercise 1

1. What does the term interest mean?
2. What is the minimum amount required to open a savings account?
3. How old must you be before you can open a cheque account?
4. What does EFTPOS stand for?
5. List three advantages of having an ATM youth account.

ISBN: 9780170217101

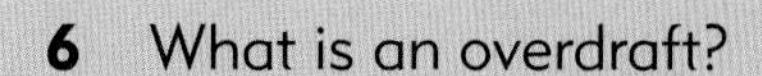

6 What is an overdraft?

7 What is a bank statement?

8 Copy the following chart and choose the description that best matches the terms in the following list:

balance credit card investment account
drawer debit card deposit slip
automatic payment cirrus

	Term	Description
a		A form used for deposits
b		The person who signs the cheque
c		A set amount deducted from your account on a regular basis
d		Worldwide interbank network
e		A money card used around the world
f		The amount of money you have in your account
g		An account with a higher rate of interest
h		Your everyday account card

ISBN: 9780170217101

Balancing up

A bank statement provides details of all your transactions and is an excellent document enabling you to monitor your money. It is very important for you to know how much money you have. The statement can be sent to you each month or at a specified time but if you have Internet banking or telephone banking capabilities, you are able to see your statement at any time. The bank is able to make electronic transactions on your behalf provided you have given them the necessary authorisation. This would involve you signing an appropriate form. These transactions can be weekly, monthly or at any specified time.

These are some of the abbreviations used on a statement:

DD: Direct debit – an amount debited from your account, it may not be the same amount each time.

AP: Automatic payment – the same amount debited from your account

DC: Direct credit – an amount of money credited to an account, it may not be the same amount each time.

A sample bank statement of transactions is provided on the following page.

ISBN: 9780170217101

Statement of: MARSH, DEBORAH LEE				
Date	**Transaction Particulars**	**Withdrawals**	**Deposits**	**Balance**
2009				
15 DEC	BALANCE BROUGHT FORWARD			14,653.43
16 DEC	INTERNATIONAL TRANSFER DEBORAH LEE MARSH PH0800863441 REFER: 09121 60006141362		2,485.00	
	CHEQUE 269998	$1,631.25		15507.18
17 DEC	NZSB ATM WITHDRAWAL KATIKATI BRANCH KATIKATI NZ	500.00		
	EFTPOS MILLS REEF TAURANGA NZ	138.80		14,868.38
21 DEC	EFTPOS COUNTDOWN CHRISTCHURCH NZ	99.29		14,769.09
22 DEC	DD AIA NZ AIA INSURE POLICY NO 91473	149.37		14,619.72
24 DEC	NON-NZSB ATM NORTHWOOD SUPA CENTA CHRISTCHURCH NZ	500.00		
	NON-NZSB ATM BALANCE ENQUIRY NORTHWOOD SUPA CENTA CHRISTCHURCH NZ			14,119.72
29 DEC	NZSB ATM WITHDRAWAL THAMES BRANCH THAMES NZ	500.00		
	EFTPOS VICTORIA PARK NEW WORLD AUCKLAND NZ Effective Dated 26 DEC 2009	56.06		
	DD LUMLEY FINANCE (N.Z. LUMLEY FIN. 3989372 INSURANCE	218.36		13,345.30
30 DEC	EFTPOS WHANGAMATA SUPERETTE WHANGAMATA NZ	71.87		
	LOAN PAYMENT MARSH DEBORAH LEE 434016173224 Effective Dated 29 DEC 2009	558.90		12,714.53
31 DEC	DC QUINVIC 5 TURRETT RD TG120166 RENT		524.54	
	DD TGA GENERAL FUNDS DL RATES 8403700D TAURANGA	135.40		
	LOAN PAYMENT MARSH DEBORAH LEE 434016175004 Effective Dated 30 DEC 2009	3,122.04		9,981.63
2010				
5 JAN	EFTPOS WILSON PARKING P66 PARK ROAD GRANZ Effective Dated 04 JAN 2010	7.00		
	EFTPOS KAI THYME CATERING WHANGAMATA NZ Effective Dated 04 JAN 2010	49.50		
	EFTPOS BP 2GO JERVOIS ROAD AUCKLAND NZ	174.25		
	EFTPOS SHIMMERS – WHANGAMATA WHANGAMATA NZ Effective Dated 02 JAN 2010	369.75		
	AP MARSH D L CARD SERVICE	300.00		
	DD FID LIFE ASS CO DL FIDELITY 98068 L5MO098778	323.06		8,758.07
	Totals at End of Page	8,904.90	3,009.54	8,758.07

ISBN: 9780170217101

To balance your cheque book with the bank statement you need to complete a **bank reconciliation**.

Use the bank statement opposite to consider the following questions.

- What is a debit?
- What is a credit?
- What does EFTPOS mean?
- What does cheque 269998 mean?
- What does NON-NZSB ATM mean?
- What does AP MARSH D L CARD SERVICE mean?
- On 29 December a withdrawal of $218.36 was made. Explain this transaction.

Reconciling your bank account

It usually takes about 24 hours for deposits or withdrawals to register on your bank statement. Sometimes a bank statement will contain transactions that are still to be cleared.

For example, if the balance in a bank statement is $246.92 but the cheque book balance is $241.77, we need to *deduct* any unpaid cheques and *add* any deposits.

Unpaid cheques	801	$64.20	
	803	$15.95	
Deduct →		$80.15	= $166.77

Deposits	$50.00	
	$25.00	
Add →	$75.00	= $241.77 (equals cheque book balance)

ISBN: 9780170217101

Exercise 2

You will find blank copies of these deposit slips and cheques at www.nelsonsecondary.co.nz/saving.

1 Imagine you are Tom Smith. You bank at the Tauranga branch of a savings bank and your account number is 01-0413-0834567-30. Today you are depositing a cheque of $645.35 from Comid Holdings, who bank with the Kawerau branch of the BNZ. You also have $250 cash to bank. Use this information to copy and complete your print out of the front and back of the deposit slip below.

Customer receipt

Date

Total amount

389813

NZSB

Deposit slip

Date

NZSB

Account holder (please print)

389813

Name of depositor (please print)

Contact phone number of depositor

Reference (limit 12 characters)

NZSB EFTPOS card number

OR Account number

0

50

$

||ᑊ50

Notes $

Coins $

Subtotal $

Cheques (as per reverse) $

Total

Cheque details

NZSB

Drawer	Bank	Branch	Amount
		Total	$

NZSB New Zealand Savings Bank Limited

2 On 25 May you purchased a computer for $1990.00 from Coastal Electronics. Using this information, copy and complete the blank cheque below.

New Zealand Savings Bank

NZSB

Account Payee

Pay

Date

Cheque No Branch Sort Code Account No. Transaction Code

729808 54 8045 44586512 01

ISBN: 9780170217101

3 You need to deposit a cheque from A. B. Smith for $94.50 (for plants) and $75 cash. A. B. Smith banks with the BNZ Katikati. Your account details are F Parent, 01-4341-0089866-00. Using this information, copy and complete this deposit slip.

Customer receipt

Date

Total amount

389813

NZSB

Deposit slip

Date

NZSB

Account holder (please print)

389813

Notes $

Coins $

Subtotal $

Cheques (as per reverse) $

Name of depositor (please print)

Contact phone number of depositor

Reference (limit 12 characters)

NZSB EFTPOS card number

OR Account number 0 50 $

Total

||' 50

Cheque details

NZSB

Drawer	Bank	Branch	Amount
		Total	$

NZSB New Zealand Savings Bank Limited

ISBN: 9780170217101

Exercise 3

Statement of: MARSH, DEBORAH LEE				
Date	**Transaction Particulars**	**Withdrawals**	**Deposits**	**Balance**
2011				
21 FEB	BALANCE BROUGHT FORWARD			2,796.91
	EFTPOS BP CONNECT TAUPO TAUPO NZ Effective Dated 20 FEB 2011	85.80		
	NON-NZSB ATM MOUNT EDEN AUCKLAND NZ	200.00		
	NON-NZSB ATM BALANCE ENQUIRY			2,511.11
23 FEB	DC CRAIGS INVESTMENT DL		212.81	
	PAYMENT/SALARY ASCOT SCHOOL PAYROLL PAYROLL 15365-5		2,105.86	4,829.78
25 FEB	DC QUINOVIC 5 TURRET RD TG1200133 RENT		261.65	
	DC CROWN RELOCATIONS CROWN RELO REFUND PAYMENT		673.38	
	CHEQUE 270030	146.87		
	CHEQUE 270033	746.75		
	DD LUMLEY FINANCE NZ LUMLEY FIN.	235.57		
	AP AK AND DG MANDENO RENT	950.00		
	DD TGA GENERAL FUNDS DL RATES 6612544370 TAURANGA	1,335.67		2,349.95
28 FEB	CHEQUE 270032	110.42		
	NZSB ATM WITHDRAWAL ST LUKES BRANCH AUCKLAND NZ Effective Dated 26 FEB 2011	200.00		
	EFTPOS FARMERS – ST LUKES AUCKLAND NZ Effective Dated 26 FEB 2011	60.00		
	LOAN PAYMENT MARSH DEBORAH LEE 434016175004 Effective Dated 30 DEC 2009	105.11		1,874.42
1 MAR	AP MARSH D L CARD SERVICE	300.00		1,574.42
2 MAR	DC QUINOVIC 5 TURRET RD TG34535 RENT		250.15	
	CHEQUE 270031	156.17		1,668,40
3 MAR	DC-4257-8949-7630 FROM 46 A/C		2,000.00	
	AP 6883-5868-0808 LOAN PMT DUE 28/2	3,122.04		546.36
4 MAR	EFTPOS PAK N SAVE ROYAL OAK ROYAL OAK NZ	90.89		455.47
7 MAR	NZSB ATM WITHDRAWAL ST LUKES BRANCH AUCKLAND NZ Effective Dated 05 MAR 2011	100.00		
	CHEQUE BOOK DUTY	1.50		353.97
9 MAR	DC QUINOVIC 5 TURRETT RD TG5680 RENT		261.65	
	PAYMENT/SALARY ASCOT SCHOOL PAYROLL/PAYROLL 14490-5		2,279.62	
	CREDIT INTEREST PAID		0.81	2,896.05
	Totals at End of Page	7,946.79	8,045.93	2,896.05

ISBN: 9780170217101

Find the following features in the bank statement on the previous page.

1 Date of the statement.
2 The value of cheque 270032.
3 Describe the transaction on 1 March.
4 How much is the cheque book duty?
5 What is the balance at 28 February?
6 What does NON-NZSB ATM mean?
7 What was the balance brought forward?
8 What does AP AK AND DG MANDENO RENT mean?

KiwiSaver

All banks offer a number of different ways to save money, so it is important for you to discuss and decide what savings plan best suits you. New Zealand has introduced a voluntary membership plan called KiwiSaver, which is primarily a retirement scheme.

KiwiSaver is a savings initiative that is supported by contributions from the government, your employer, and yourself. Savings are built up through regular deductions from your pay that are made by your employer. These contributions can vary from 2%, 4% or 8% of your pay.

When you first join KiwiSaver the government will contribute $1000, which is also called a kickstart. Money is generally locked in until you are eligible for retirement and/or you have been a member for at least five years. Money can be withdrawn early, but only if you are either:

- Buying your first home.
- Moving overseas permanently.
- Suffering sufficient financial hardship.
- Seriously ill.

ISBN: 9780170217101

Exercise 4

1 Oliver earns $560.50 a week. How much would his contributions be at 2%, 4% and 8%?

2 Mary earns $497.00 a week. If she has been contributing 4% of her weekly pay for five years …

- **a** How much has she saved in her KiwiSaver account?
- **b** If her employer has been contributing 4% as well, how much has her employer contributed?
- **c** How much does she have in her KiwiSaver account after five years?

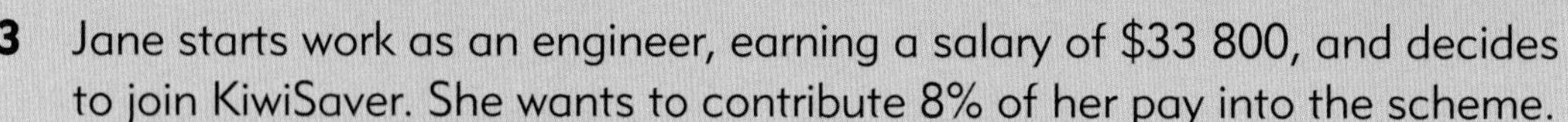

3 Jane starts work as an engineer, earning a salary of $33 800, and decides to join KiwiSaver. She wants to contribute 8% of her pay into the scheme.

- **a** How much is the kickstart?
- **b** How much is her weekly contribution?
- **c** How much is her weekly employer contribution?
- **d** How much does she have in her KiwiSaver account at the end of the first year?
- **e** How much does she have in her account at the end of ten years? (Assuming her salary remains the same.)

ISBN: 9780170217101

Reconciliation in more detail

To reconcile or balance the cheque book and bank statement given on the next page, follow these steps.

Steps	Example	
1 Bank statement balance Cheque book balance		$430.00 $384.55
2 Mark off cheque and deposit transactions from the cheque book on to the bank statement.		
3 Total any unpresented cheques and deduct this amount from the bank statement balance.	806: 808:	$55.00 $76.95 $131.95 $430.00 – $131.95 = $298.05
4 Total any deposits and add this amount to the bank statement balance.		$80.00 $298.05 = $378.05
5 Adjust the cheque book balance by deducting any automatic debits or credits shown on the statement, that are not recorded by you.	 Fee Ch. bk	$384.55 $5.00 $1.50 – $6.50 = $378.05
6 Bank statement balance = cheque book balance		$378.05

ISBN: 9780170217101

STATEMENT OF ACCOUNT WITH
Name Mr R Hoodwink
Street 25 Violet Lane
Town Tauranga

Date	Particulars	Debit	Credit	Balance
Feb 6 '12	Balance forward			454.86 Cr
Mch 4 '12	804	15.86		439.00 Cr
Mch 5 '12	805	500.00		61.00 OD
Mch 10 '12 C/C			500.00	439.00 Cr
Mch 10 '12	807	2.50		436.50 Cr
Mch 10 '12 FEE		5.00		431.50 Cr
Mch 10 '12 CBK		1.50		430.00 Cr

EXPLANATION OF ABBREVIATIONS USED

C/C means	Cash and/or Cheque lodged for credit	FEE means	Charge for keeping Account
CBK -	Cheque Book	INT -	Interest on Account
DIV -	Dividend	TFR -	Transfers from other Banks or Branches
CTI -	Interest on Commonwealth Treasury	OD -	Overdrawn

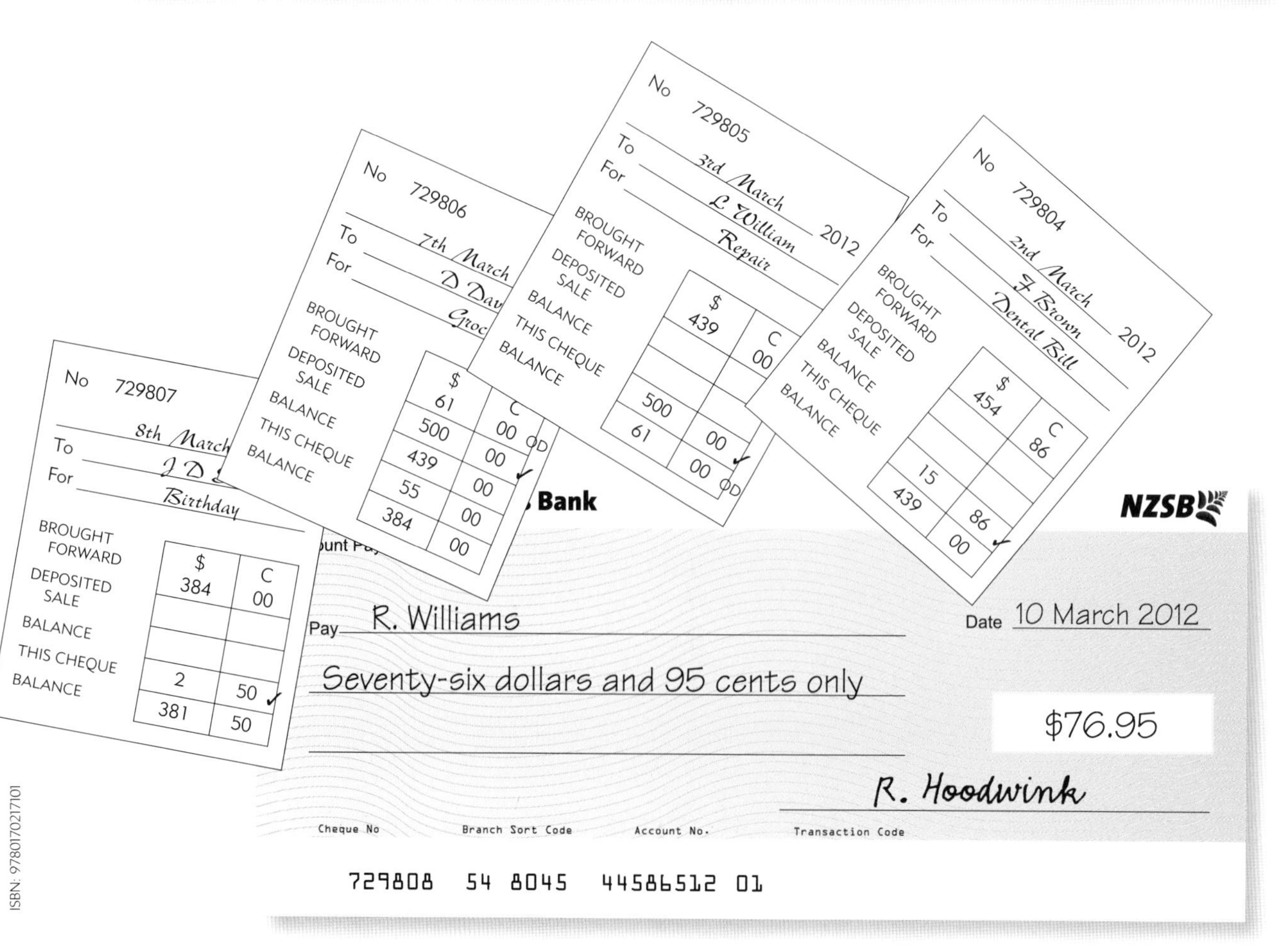

ISBN: 9780170217101

Exercise 5

1 Copy and complete the balance column of this table.

Date	Withdrawal	Deposit	Balance
18-7-12		200.00	424.55
20-7-12	120.00		
21-7-12		150.00	
25-7-12	84.00		
26-7-12	37.95		
29-7-12	24.75		
30-7-12		350.00	
30-7-12	112.15		
31-7-12		200.00	
01-8-12	192.45		

2 Answer the following questions using the bank statement on the next page.

a What type of account is it?

b Whose account is it?

c What is the balance brought forward? What does this mean?

d What is the starting date of the statement?

e What was the balance on 29 October?

f Explain the three transactions that took place on 1 November.

g What is the total amount of withdrawals for the period of the statement?

h What is the closing balance?

ISBN: 9780170217101

Statement of: MARSH, DEBORAH LEE

NZSB Access Account

Date	Transaction Particulars	Withdrawals	Deposits	Balance
2010				
14 OCT	BALANCE BROUGHT FORWARD			17,075.49
15 OCT	DC WILLIS NZ CROWN WW 1202-03148-03		1,385.40	
	CHEQUE 27011	138.14		
	LOAN PAYMENT MARSH DEBORAH LEE 43413581523 Effective Dated 14 OCT 2010	3,559.96		14,762.79
18 OCT	EFTPOS DFS AK AIRPORT AIR 81 AUCKLAND NZ Effective Dated 14 OCT 2010	69.00		14,693.79
20 OCT	PAYMENT/SALARY ASCOT SCHOOL PAYROLL PAYROLL 13429-50909		2,240.92	
	DD AIA NZ 94654 95541331H	170.40		16,764.31
21 OCT	DC QUINOVIC 5 TURRETT RD TG430173 RENT		523.30	
	CHEQUE 270012	240.61		17,047.00
22 OCT	DC CRAIG & CO CASH DL CRAIGS IP CASH MGMT TRUST LTD		3,476.06	
	AP AK AND DG MANDENO RENT	950.00		19,573.06
28 OCT	INTERNATIONAL TRANSFER WESTERN ACADEMY OF B PH023547245 REFER: 1012 8007551829		3,269.57	22,842.63
29 OCT	DC GRIFFINS FOO 1500114080		486.85	23,329.48
1 NOV	EFTPOS KC LOO FRUIT SHOT MOUNT EDEN NZ Effective Dated 31 OCT 2010	55.00		
	AP MARSH D L CARD SERVICE	300.00		
	NON-NZSB ATM MOUNT EDEN AUCKLAND NZ Effective Dated 31 OCT 2010	200.00		22,774.48
2 NOV	LOAN PAYMENT MARSH DEBORAH LEE 43461001941823 Effective Dated 01 NOV 2010	3,122.04		19,652.44
3 NOV	PAYMENT/SALARY ASCOT SCHOOL PAYROLL PAYROLL 1352-5		539.59	
	NZSB INTERNET FUND TRANSFER TO 4988730050011550 276523	1,000.00		19,192.03
4 NOV	DG QUINOVIC 5 TURRETT RD TG120154 RENT		511.80	
	DC PUBLIC TRUST ESTATE OF E B HUNTER PUBLIC TRUST		7,375.71	
	CHEQUE 270013	2,137.39		
	EFTPOS THAI EDEN MT EDEN AUCKLAND NZ	66.00		24,846.15
5 NOV	DC PUBLIC TRUST ESTATE OF E B HUNTER PUBLIC TRUST		65.68	
	AP AK AND DG MANDENO RENT	950.00		23,991.83
	Totals at End of Page	12,958.54	19,874.88	23,991.83

ISBN: 9780170217101

Making a purchase

There are two ways of making a purchase: by cash or by credit.

Cash buying

The goods are paid for by EFTPOS, cash or a cheque at the time of purchase.

Advantages		Disadvantages	
1	You could get a discount on the asking price.	**1**	The risk of carrying around cash.
2	You can take the goods away and use them immediately.	**2**	You may miss out on a good buy because you don't have the cash available at the time.
3	You only buy when you can afford it.		
4	Money is accepted everywhere.		
5	You stay out of debt.		

Lay-by is another form of cash buying. Goods are put away and paid for by instalments of an agreed amount between the purchaser and shop owner. The purchaser only pays the cash price, and will not get possession of the goods until the final payment is made. Fashion goods are the most common items sold on lay-by terms.

ISBN: 9780170217101

Exercise 6

1 Jane wished to purchase an iPod marked at $499. She was offered a 15% discount if she paid cash. What was the discount and her purchase price?

2 Stan was prepared to pay cash for a pink Volkswagen priced at $8500. He was offered an $800 discount for cash. What percentage is this of the marked price? (Answer to 1 d.p.)

3 The Wilson family of four were going on a holiday to Surfers Paradise for ten days in the August holidays. Their travel and accommodation costs were $4320, but if they paid it one month in advance they could have a 12.5% discount. If they paid two months in advance they could have an 18% discount.

- a What was the 12.5% discount?
- b What is the difference (in dollar terms) between the 18% and 12.5% discount?
- c Which would you choose? Why?

Credit buying

When purchases are made on credit, you are agreeing to pay for them at a later date.

	Advantages		Disadvantages
1	You can budget for the payments.	1	You pay interest on the credit.
2	Have the use of the goods while it is being paid for.	2	Can encourage impulse buying.
3	No need to carry around large amounts of money.	3	Can commit yourself to too many purchases.
4	Allows you to buy goods at special prices.	4	False sense of security.
5	Convenience.	5	Can make savings difficult.
6	Can meet unexpected costs.		

There are three different methods of making a purchase by using credit:

- Monthly account.
- Hire purchase.
- Credit card.

ISBN: 9780170217101

1 Monthly accounts

Nowadays most purchases are paid for by EFTPOS, cash or a credit card. However, businesses still use monthly accounts whenever purchases are being made regularly or on a daily basis.

An invoice giving specific details of the purchase is issued at the time of sale, and then at the end of each month a statement of account is sent to the purchaser, which details all the invoice numbers for that period.

The purchaser reconciles their invoice number with the statement number and the amount charged.

2 Hire purchase

Many people use hire purchase (HP) to buy large value items such as a car, computer or household appliances. It is a very useful method when used sensibly, because you may not have enough ready cash at the time to pay the full amount for the goods. It is a type of loan.

In order to make a hire purchase you enter into a contract with the supplier and pay a small deposit straight away, as well as agreeing to pay the balance, plus interest, in instalments. You have the use of the goods while they are being paid for.

The goods are still owned by the store, however, until payment is completed. If the purchaser fails to make their HP payments, the goods can be repossessed. Purchasers must sign an HP agreement at the beginning of the contract, so you need to be aware of what you are signing.

Exercise 7

1 Copy and complete this table.

	Item (with cash price)	Deposit	Instalments	Total price	Interest (extra paid)
	Mower, $495	10% = $49.50	12 instalments @ $42 = $504	$504 + $49.50 = $553.50	$553.50 - $495 = $58.50
a	Bike, $995	15%	18 instalments @ $62		
b	iPod, $495	20%	10 instalments @ $58.50		
c	Computer, $1995	10%	24 instalments @ $85.25		

ISBN: 9780170217101

	Item (with cash price)	Deposit	Instalments	Total price	Interest (extra paid)
d	Mini (car), $14 995	10%	36 instalments @ $395.00		
e	P class yacht, $2560	25%	24 instalments @ $99.95		

To calculate **simple interest** or **flat interest**, the formula is:

$$\text{S.I.} = \frac{PTR}{100}$$

where S.I. = simple interest, **P = principal** (amount of money invested), **T = time in years**, and **R = rate**.

2 Using the formula above, calculate the S.I. on:

a $5000 invested for two years at 6.5%

b $8000 borrowed for three years at 9.5%

c $5500 borrowed for two years at 10.5%

d $1000 borrowed for 18 months at 8.75%

e $750 borrowed for six months at 11.75%

3 Rearrange the formula S.I. = P.T.R. / 100 to make:

a P the subject of the formula.

b T the subject of the formula.

c R the subject of the formula.

4 Using the formulae from question 3 calculate:

a R if $4000 is invested for two years and earns $250 interest.

b R if $2500 is invested for six months and earns $50 interest.

c P if $400 interest is earned in five years at 10%

d P if $300 interest is earned in two years at 7.5%

e T if $3000 is borrowed at 10% and earns $150 interest.

f T if $5000 is borrowed at 5.5% and earns $1168.75 interest.

5 Tony borrows $10 000 from his bank and arranges to pay it back at $368 per month over three years:

a How much in total does he repay?

b How much interest does he pay?

c What was the flat interest rate? (1 d.p.)

ISBN: 9780170217101

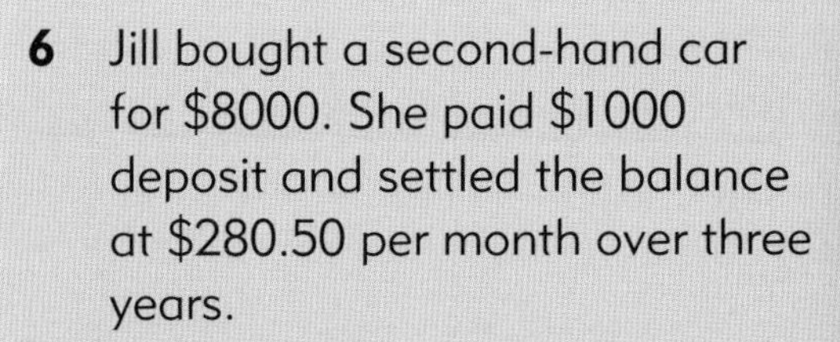

6 Jill bought a second-hand car for $8000. She paid $1000 deposit and settled the balance at $280.50 per month over three years.

- **a** How much were the repayments in total?
- **b** How much did the car cost in total?
- **c** How much interest did she pay?
- **d** What percent is this interest of the cost price of the car? (1 d.p.)

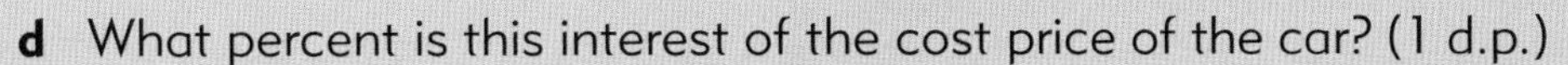

7 Ashley bought a tent for $1050 and paid 25% deposit. The balance was paid in 12 instalments of $86.85.

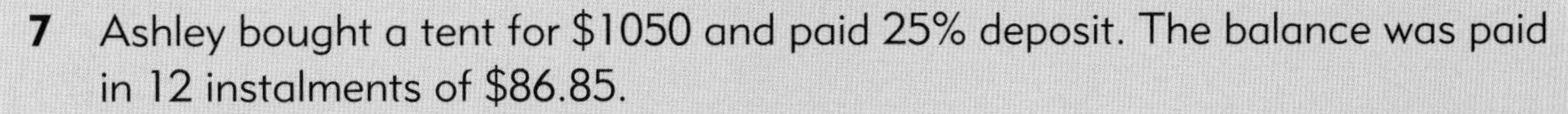

- **a** What was his deposit?
- **b** How much did his instalments total?
- **c** What was the total cost of the tent?

8 John bought a second-hand car for $6500. He agreed to pay a deposit of $1000, then $200 per month for three years and $250 per year for insurance.

- **a** At the end of three years, how much would the car have cost him?

  After six payments he found the car too expensive. The finance company repossessed and sold it for $4000. The finance company then sent him an account for $1022.50 for the balance of money owing on the hire purchase agreement. He paid the account and had no car.

- **b** How much had he paid out altogether after six months for the car he didn't have?

ISBN: 9780170217101

Compound interest

This occurs when the interest earned from the investment is added or subtracted from the principal at regularly decreasing intervals, so that the principal is always changing.

For example, if $1000 is borrowed at 12% for 18 months compounding six monthly:

1st 6 months:

$$\text{S.I.} = \frac{\text{P.T.R.}}{100}$$

$$= \frac{1000 \times 0.5 \times 12}{100}$$

$$= \$60$$

$$\text{Principal} = \$1000 + \$60$$

$$= \$1060$$

2nd 6 months:

$$\text{S.I.} = \frac{1060 \times 0.5 \times 12}{100}$$

$$= \$63.60$$

$$\text{Principal} = \$1060 + \$63.60$$

$$= \$1123.60$$

3rd 6 months:

$$\text{S.I.} = \frac{1123.60 \times 0.5 \times 12}{100}$$

$$= \$67.42$$

$$\text{Principal} = \$1191.01$$

$$\text{Interest paid over 18 months} = \$1191.01 - \$1000$$

$$= \$191.01$$

Exercise 8

1 Using the above method, calculate the interest on $6000 at 8.5% compounding quarterly for one year.

2 Steve has to borrow $5000 for two years to buy a new car. He is given two choices:

- i A flat interest rate of 11.5%
- ii An interest rate of 9.5% compounding every six months.

ISBN: 9780170217101

a How much interest is paid in i?
b How much interest is paid in ii?
c Which is the better choice for Steve?

Compound Interest can also be calculated using this formula:

$$FV = PV \times (1 + \frac{r}{100})^n$$

FV is the full (final) value of the investment/borrowing
PV is the present value of the investment/borrowing
r is the rate of interest per annum
n is the number of compounding interest time periods

eg Calculate the FV and Interest earned if $1000 is invested for five years, compounding annually at 5.5% pa.

$$FV = 1000 \times (1 + \frac{5.5}{100})5$$

FV = $1306.96

Interest = $306.96

Use this formula to answer questions 3, 4 and 5.

3 Calculate the full value of the following investments (answer to nearest $).
a $2000 invested for three years @ 6% pa compounding annually.
b $5000 invested for five years @ 6.5% compounding annually.
c $10 000 invested for four years @ 5% compounding every 6 months.
d $8500 invested for two years @ 4.25% compounding every month.

4 Calculate the compound interest you will need to pay on the following borrowings.
a $3000 borrowed for two years @ 8% compounding annually.
b $10 000 borrowed for three years @ 7.5% compounding annually.
c $5000 borrowed for five years @ 6.25% compounding every 6 months.
d $2500 borrowed for one year @ 8.25% compounding monthly.

5 Calculate each of these options and decide which is the best option for you if you need to borrow $5000 to buy your first car.
Option A: Pay simple interest over four years @ 6.75%
Option B: Pay compound interest over three years @ 6.85% compounding annually
Option C: Pay simple interest over six years @ 4.25%.

ISBN: 9780170217101

3 Credit cards

A credit card is a safe and convenient way to buy almost anything. They also give you access to cash advances from ATMs anywhere in New Zealand or overseas. All the different banking groups have different card names, however, the basic services are the same.

You must be over 18 years of age and have a satisfactory banking record with the bank before you will be issued with a credit card.

Whenever you make a purchase using your card you will be given a sales voucher to sign. The signature on the card and voucher is always checked to make sure they are the same. You then get the goods to use. (The store is paid by the finance company supplying the card, and you pay the finance company later.) Every month each cardholder is issued with a monthly statement detailing all their purchases.

If you pay the full amount within 25 days of the date of your statement then no interest is charged. However, if you do not wish to pay this full amount you are required to make the minimum repayment shown on the statement. In this situation you will be charged interest on all outstanding advances at a daily compounding interest rate (currently 22.2% pa).

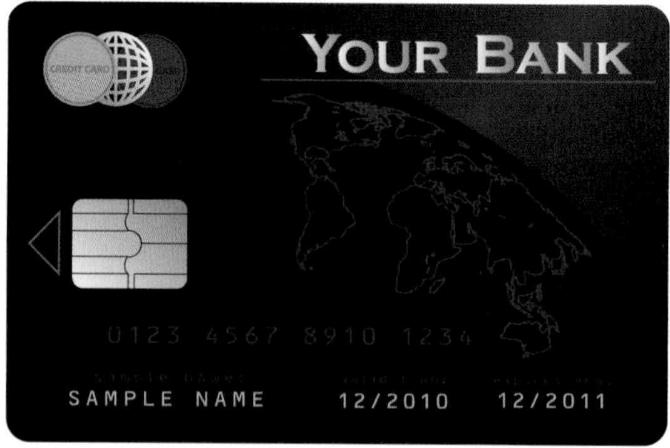

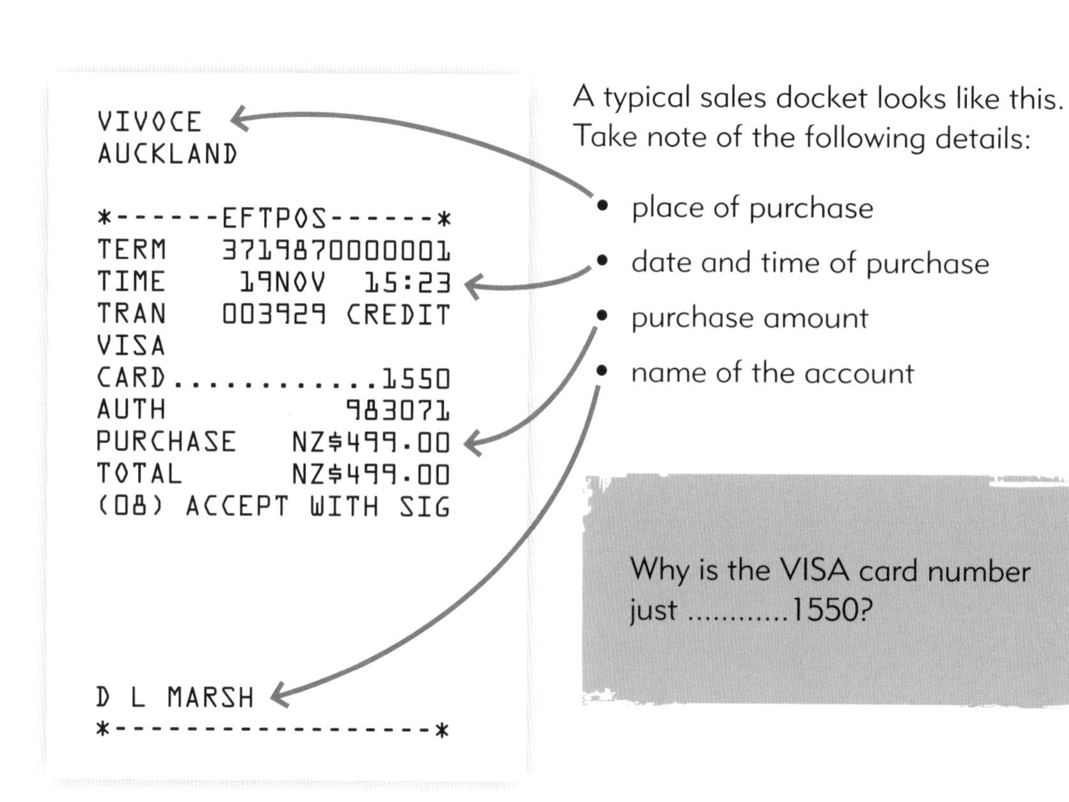

VIVOCE
AUCKLAND

------EFTPOS------
TERM 3719870000001
TIME 19NOV 15:23
TRAN 003929 CREDIT
VISA
CARD............1550
AUTH 983071
PURCHASE NZ$499.00
TOTAL NZ$499.00
(08) ACCEPT WITH SIG

D L MARSH

A typical sales docket looks like this. Take note of the following details:

- place of purchase
- date and time of purchase
- purchase amount
- name of the account

Why is the VISA card number just ....1550?

ISBN: 9780170217101

Exercise 9

Use the credit card statement that follows to answer the questions on the next page.

MRS DEBORAH L MARSH
Flat 2
328 Mount Eden Rd
Mount Eden
Auckland 1024

Customer service freephone	0800 658 044
Statement period	13/12/2010 to 10/01/2011
Your credit limit	$51,000.00
Your available credit	$49,625.57

Interest rates

PURCHASES	INTEREST RATE 19.95% p.a.
CASH ADVANCES	INTEREST RATE 22.20% p.a.
BALANCE TRANSFERS	INTEREST RATE 19.95% p.a.

New Zealand Savings Bank
NZSB

Transaction details

Date processed	Date of transaction	Card used	Transaction details/Currency	Origin/Effective exchange rate	Amount (NZ$)
15/12/2010	11/12/2010	1550	ESETEE LAUDER 170.00 AUD INCL CURRENCY CONVERSION CHARGE	WOOLLAHRA 0.7587 $5.60	229.68
15/12/2010	14/12/2010	1550	TIMBERLAND 80.00 AUD INCL CURRENCY CONVERSION CHARGE	DRUMMOYNE 0.7587 $2.64	108.09
17/12/2010	14/12/2010	1550	WORLD KITCHEN AUST 35.15 AUD INCL CURRENCY CONVERSION CHARGE	DRUMMOYNE 0.7487 $1.17	48.12
20/12/2010	17/12/2010	1550	TOPLINE/NATUREBEE	AUCKLAND	61.50
31/12/2010	29/12/2010	1550	SHIMMERS - WHANGAMATA	WHANGAMATA	235.00
05/01/2011	05/01/2011	1550	**AUTOREPAYMENT - THANK YOU**		300.00 CR
07/01/2011	05/01/2011	1550	CCS DISABILITY ACTION INC	AUCKLAND	15.00
10/01/2011	10/01/2011	1550	ANNUAL ACCT FEE		80.00
10/01/2011	10/01/2011	1550	INTEREST CHARGED ON PURCHASES		21.53

YOU CAN REPORT YOUR CREDIT CARD LOST OR STOLEN 24 HOURS A DAY, EVEN ON WEEKENDS, BY CALLING 0800 103 123.

Drawer	Bank	Branch	Amount
			$
			$
			$

Teller Stamp

Signature

Subtotal	$
Notes	$
Coins	$
Total	$

ISBN: 9780170217101

1 What is the interest rate per annum for purchases?

2 What is the interest rate per annum for cash advances?

3 What is a cash advance?

4 What was the currency exchange rate for the purchase at Timberland on 14/12/10?

5 How much was the currency charge for the purchase above?

6 The purchase on 17/12/10 took how many days to be processed?

7 How much interest was charged on purchases?

Credit cardholder responsibilities

Being a cardholder has benefits as well as obligations.

Advantages		Disadvantages	
1	It is safe and convenient.	**1**	You might overspend.
2	You can buy now and pay later.	**2**	You must pay interest on the amount owing after 25 days.
3	You can take advantage of special prices if paid for within 25 days.	**3**	The interest rate is usually much higher than other services offered at the bank.
4	There is free credit for 25 days.	**4**	You may get a false sense of security.
5	You can access cash advances.	**5**	You are tempted to buy on impulse.
6	There is extended credit should you need it.		
7	You get a record of your transactions.		
8	You can use this for identification.		

Exercise 10

1 Find out the names of the credit card services available in your local area.

2 Find out the current annual interest rate being charged on a credit card account.

3 Answer the following questions using the credit card statement on the next page.

ISBN: 9780170217101

a What is the closing balance?

b What is the minimum payment due?

c How much interest has been charged?

d When is the payment due?

e What is the amount charged for fees and other charges?

f Were there any cash advances?

g What was the total of this month's purchases?

MRS DEBORAH L MARSH
Flat 2
328 Mount Eden Rd
Mount Eden
Auckland 1024

Customer service freephone	0800 658 044
Statement period	13/12/2010 to 10/01/2011
Your credit limit	$51,000.000
Your available credit	$49,625.57

New Zealand Savings Bank
NZSB

Primary summary	
Minimum payment due	$28.00
DUE DATE	04/02/2011

Account summary	
Opening balance	$875.51
Payments and other credits	$300.00
Purchases*	$697.39
Cash advances*	$0.00
Balance transfers	$0.00
Interest charges	$21.53
Fees and other charges	$80.00
Closing balance	$1,374.43

*Includes currency conversion charges on any foreign currency transactions - see transaction details on statement for breakdown.

Payment record	**Date paid:**	**Amount paid:**
		$

Credit card payment advice

For your convenience, we have the following payment options:
Read on for more payment options.
Direct debit.
Internet
Phone direct.
By post.

Acconut number	XXXX-XXXX-XXXX-XXXX
Account name	MRS DEBORAH L MARSH
Date paid	
Amount paid	$

4 Jane buys a $1500 sound system on her credit card. She takes four months to pay for it ($425 a month). If she gets charged 22% pa each month for the balance owing, how much credit card interest did she pay? Discuss whether this is a good buy or not.

ISBN: 9780170217101

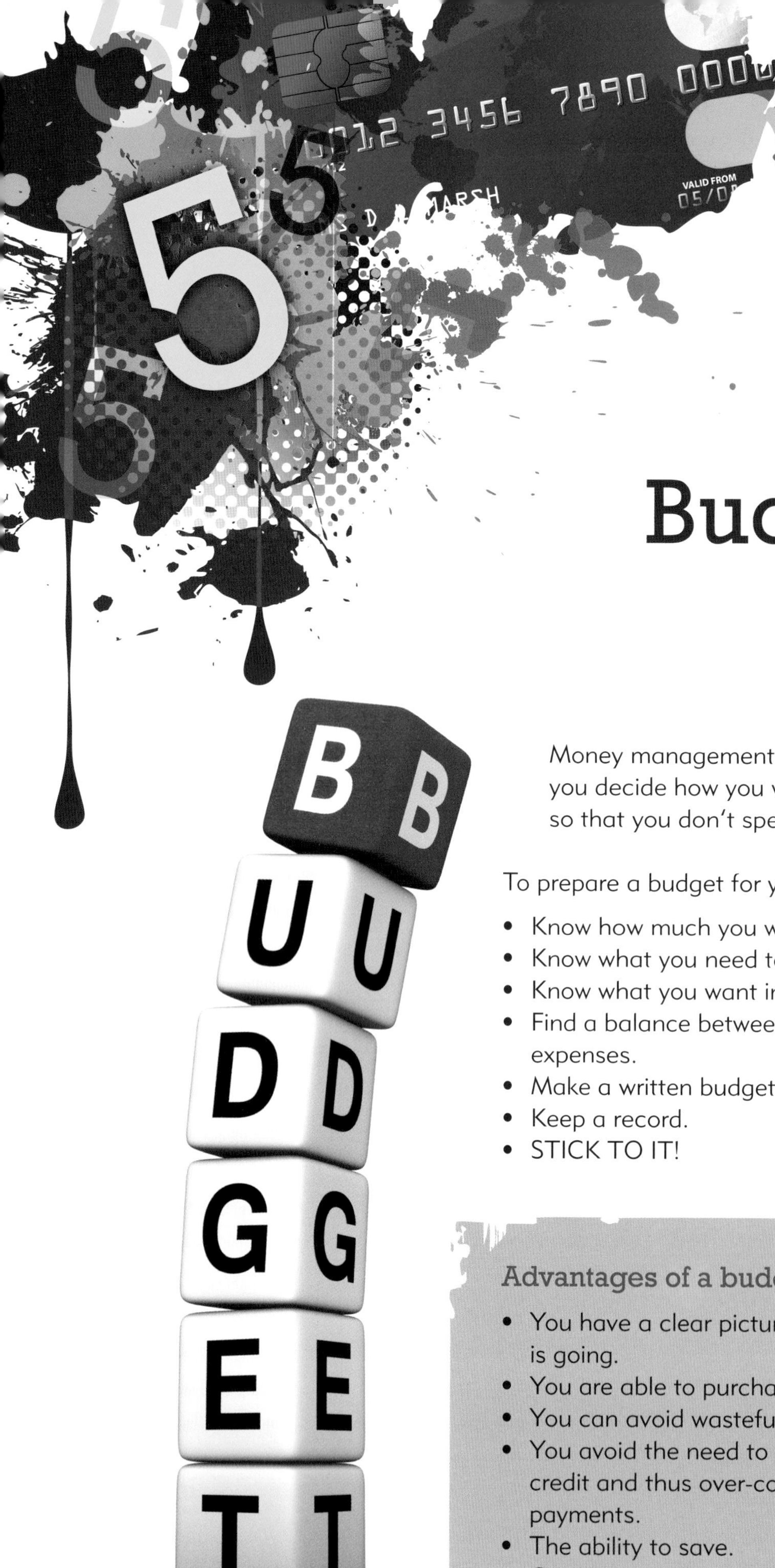

Budgeting

Money management and good planning helps you decide how you will spend your money, so that you don't spend more than you earn.

To prepare a budget for yourself:

- Know how much you will earn **(your income)**.
- Know what you need to live on **(your expenses)**.
- Know what you want in life **(your goals)**.
- Find a balance between fixed and variable expenses.
- Make a written budget.
- Keep a record.
- STICK TO IT!

Advantages of a budget

- You have a clear picture of where your money is going.
- You are able to purchase priority items.
- You can avoid wasteful spending.
- You avoid the need to borrow, over-extend your credit and thus over-commit yourself to HP payments.
- The ability to save.
- Gives you a sense of pride in being able to manage your own affairs.

ISBN: 9780170217101

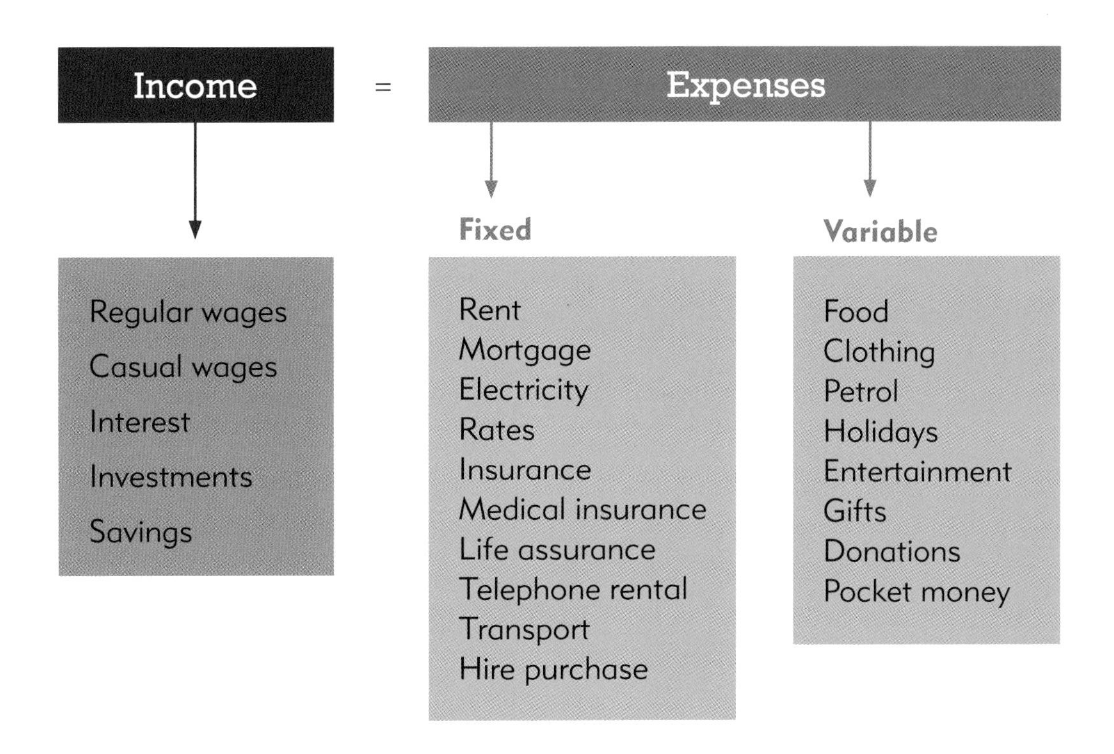

> **Remember:** Budgets are very personal because they depend on where and what your priorities are.

Exercise 11

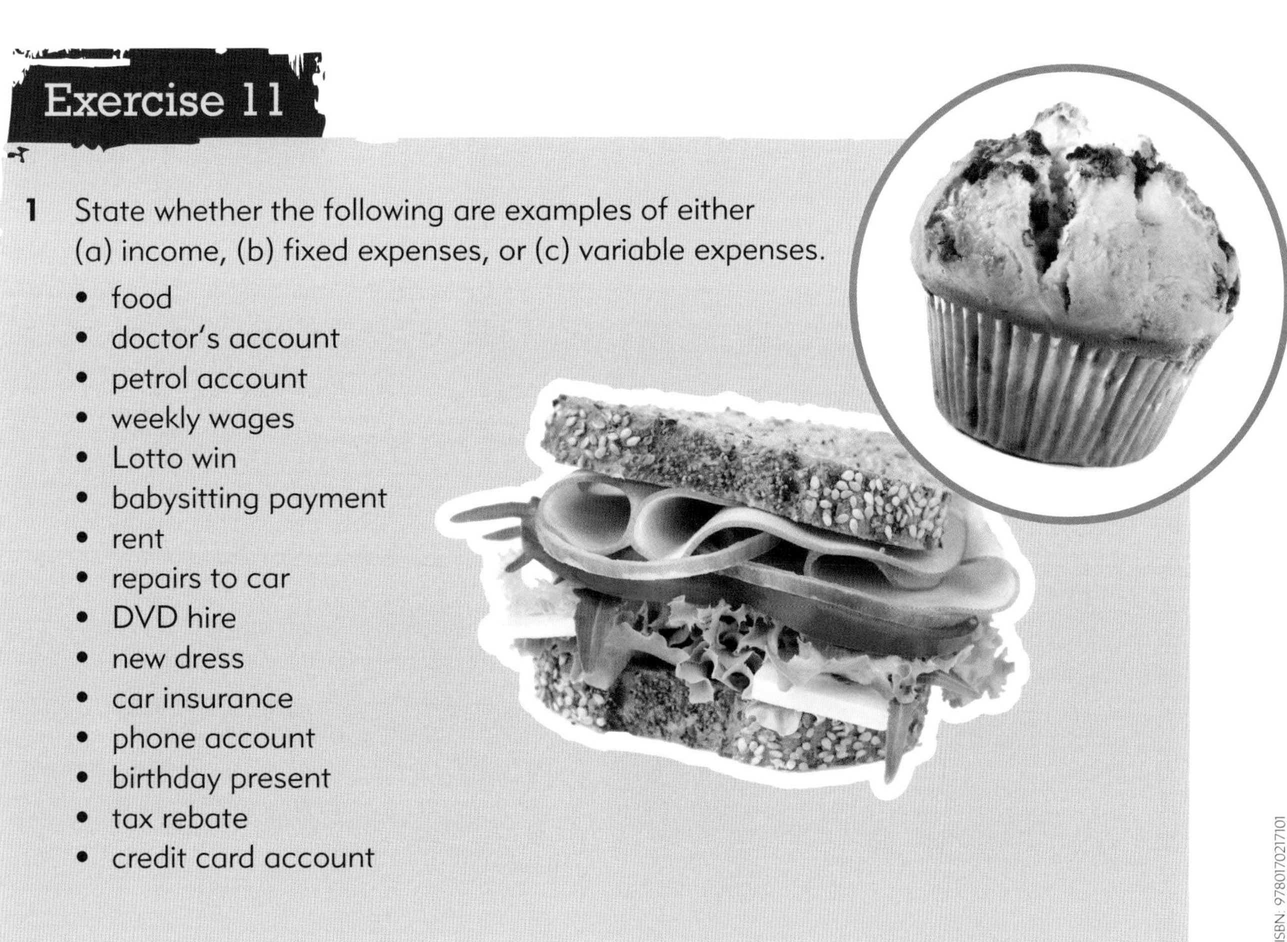

1 State whether the following are examples of either (a) income, (b) fixed expenses, or (c) variable expenses.

- food
- doctor's account
- petrol account
- weekly wages
- Lotto win
- babysitting payment
- rent
- repairs to car
- DVD hire
- new dress
- car insurance
- phone account
- birthday present
- tax rebate
- credit card account

ISBN: 9780170217101

2 With your weekly allowance of $40 you must buy lunch at school twice a week (this costs $4.50 each time), and you also have to buy your school supplies (which average $10.25 per month).

- **a** How much is your monthly income?
- **b** What is your monthly lunch total?
- **c** What other monthly expenses could you have? Give an approximate value of these.
- **d** How much money do you have left each month?
- **e** How much per week is this? (Answer to nearest five cents.)
- **f** How much per school day is this? (Answer to the nearest cent.)

3 Leanne earns $495 a week and shares a flat with two friends. She pays $65 a week for rent. She spends $50 a week on food and $35 a week on transport. Leanne's weekly share of the electricity bill is $12.50.

- **a** How much are her weekly expenses?

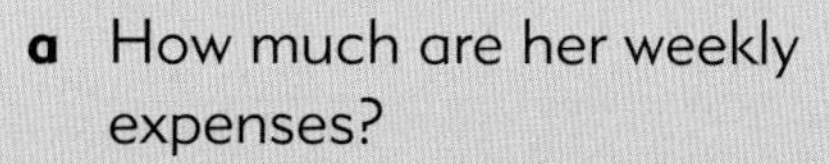

- **b** How much does she have left after paying these?
- **c** Name three other expenses she may have.
- **d** If she saves $20 a week how much will she have in 12 months?

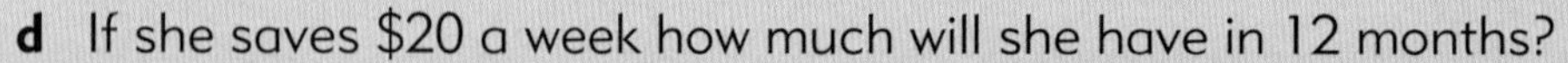

- **e** She smokes approximately two packets of cigarettes a week at $13.10 a packet. How much would she save in 12 months if she gave up smoking?

4 The Smith's monthly income is $2600. Some of their expenses for the month are:

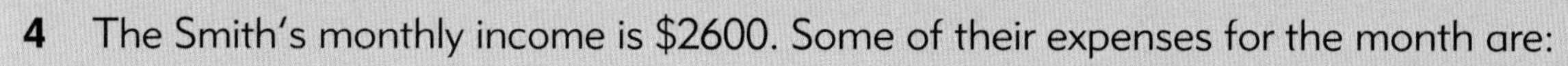

- Mortgage repayment $940
- Power account $176
- Housekeeping $700
- Petrol $100
- Doctor's account $58.50
- HP on TV $56
- Telephone and mobile phone account $172
- Pocket money $150.

ISBN: 9780170217101

a What is the total of these expenses for the month?

b Name two other expenses they may need to account for.

c Do you think the Smiths have enough money left over for unexpected expenses? Why?

d Do you think the Smiths need a savings plan? Why?

5 Next door to the Smiths live the Willburys, whose monthly income is $2800. The Wilbury's monthly expenses are:

- Rent $920
- Fares to and from work $180
- HP on TV $56
- HP on computer $59
- Car repairs $84
- Petrol $122
- Housekeeping $700
- Pocket money $200.

a What is their total monthly expenses?

b How much do they have left each month?

c Which family has the better budget?

d What adjustments would you make, and why?

6 When you go flatting you will have lots of expenses other than rent. List these expenses and their likely costs. If your share of the rent is $120, how much will you need to earn to cover these costs?

7 Make up your own monthly budget based on your income and expenses.

8 Name ways of increasing your income.

9 Name ways of reducing your expenditure. Explain.

10 Select something you would like to save for. Find out how much it will cost and how you will adapt your budget to include this item.

ISBN: 9780170217101

Setting up a business

If you are thinking of being self-employed, the first thing you need to do after a feasibility study, is to prepare a budget. These should be part of your business plan. The feasibility study should look at the strengths and weaknesses of existing businesses in your chosen area and whether from your research, setting up a similar business is a sound, viable idea. A budget should outline your predicted income and expenses. Expenses will cover such things as rent, telephones, power, marketing/ advertising, insurances, wages, vehicles, materials and/or tools needed for the business. If you are borrowing money from the bank or require an overdraft, your bank will require you to complete this kind of business plan.

Exercise 12

1 Leasing the correct building location can be a significant factor in the success of the business. Calculate the cost per square metre per week of the following buildings on offer from a Real Estate Agent.

a Option A – 300 square metres, $3150/month

b Option B – 350 square metres, $58 500/yr

c Option C – 420 square metres, $4450/month

d Option D – 380 square metres, $825/week

Which option would you choose? Why?

2 What are some of the characteristics of a location you would need to consider if you were planning to develop the following businesses?

a Hairdressing

b Jewellery manufacturing

c Small delivery truck business

ISBN: 9780170217101

d Graphic design

e Making cup cakes

f Café

3 You have leased a 400 square metre space for a café. It is 40 m long and 10 m wide which includes the footpath frontage. Prepare to scale the layout of your café.

4 When you buy goods from a wholesaler or directly from the grower, you do need to add some mark-up to the goods. Calculate the percentage mark-up on the following goods.

a Herbs bought for $0.90/bunch, sold for $3.95/bunch.

b T-Shirts bought for $100/50 shirts, sold for $15.95 each.

c Fruit bought for $0.75/kg, pulped and sold as pulp for $10.95/L

5 When you are in business, GST (15%) needs to be added to your goods and services and a record of this calculation needs to be kept. Each month two months/ six months or year, you need to calculate the GST received with the GST paid on goods and services. The difference is paid to the IRD. Calculate the GST payable each month on the following goods purchased for your business (prices inc GST).

a Hairdressing services of $4050 a month.

b Printing goods to the value of $2595 used each week.

c Truck hireage to the value of $10 500 a month.

d $125.00 of petrol used each week.

6 Prepare a monthly Profit and Loss Statement for the following business – discuss your findings.

Sarah's Laundry Services

Item	$	c	Frequency
Income	$6590	00	per week
Advertising	$350	00	per week
Lease	$840	00	per month
Electricity	$425	00	per month
Telephone	$395	00	per month
Laundry chemicals	$425	00	per month
Wages	$1010	00	per week
Insurance	$85	00	per month
Rates	$195	00	per week
GST	$650	00	per week
Taxes	$99	90	per week
Vehicle use	$95	00	per week

ISBN: 9780170217101

7 You plan to buy a hairdressing salon. Prepare a budget, outlining income and expenses you will need to cover each month.

8 You are planning to buy your own truck and contract to the local building merchant to deliver goods from the warehouse to building sites and private residences. Prepare a budget, outlining expected income and expenses you will need to cover each month.

9 Investigate what small business opportunities are available in your local community.

10 Visit your local bank and see what opportunities are available for small business developments in your local community.

11 Investigate the Young Enterprise Businesses of the Year and see what has worked and how they have been successful.

ISBN: 9780170217101

Getting the best buy

Everyone likes a bargain. This is why advertisers so frequently use statements like these.

To help ensure you get the best buy, you should ask yourself these questions:

- Do I really need this item?
- Do I really need three or five of them, or will one be enough?
- Is it really the best price available?
- Is it worth the price?
- Does this shop guarantee its goods?
- Are these good quality goods?
- What is the usual price?

ISBN: 9780170217101

Exercise 13

1. Collect some advertisements that encourage you to make a purchase at a special price.

2. Which is the best value?
   a. Coffee 50 g @ $3.53 or 100 g @ $5.85 or 1 kg @ $27.20?
   b. Spaghetti 300 g @ $0.90 or 430 g @ $1.10 or 830 g @ $2.05?
   c. Laundry powder 400 g @ $4.40 or 1 kg @ $9.95?
   d. Peaches 300 g @ $1.05 or 425 g @ $1.40 or 830 g @ $2.85?
   e. Stockings one pair @ $2.99 or pack of three pairs @ $8.99?
   f. Cola 375 ml @ $1.40 or 1.5 L @ $5.95?
   g. Tissues pkt 180 @ $1.95 or pkt 200 @ $2.05?

3. Milk can be purchased at a range of quantities and from a variety of different outlets. Find out what these are and the relative prices. For a flat of four people, which would be the best purchase?

4. Potatoes are advertised at 20 kg @ $6.95, 5 kg @ $2.10 and 10 kg @ $4. There are only two of you in the flat. Discuss which would be the better buy? Why?

5. Work socks are advertised at either $5.99 a pair or a pack of three pairs for $12. What is the saving in buying the pack of three?

6. Jane bought a new tyre for $110, Sarah bought one for $85, and Annabel a retread for $60. Jane's tyre lasted 40 000 km, Sarahs 30 000 km and Annabels 20 000 km. Which was the best buy?

7. Select six grocery items of the same weight or amount and compare the prices of at least two different brands. Choose, for example: cheese, yoghurt, tinned peaches, honey, jam, dishwashing liquid, biscuits, sultanas, ice-cream.

8. Select two grocery retail outlets in your area. Make a list of at least 20 basic grocery items and compare the prices of these items between the stores. (Remember to compare products of the same size.)

ISBN: 9780170217101

Giving change

When we give change or are given change it is important to give it in the most convenient way. This means using as few coins and notes as possible.

Exercise 14

1 Using the New Zealand coin currency, how many ways can you make one dollar?

2 Using only New Zealand note currency, how many ways can you make $20?

3 Susan makes the following purchases and each time hands over $10. What is the most convenient way of giving her change?

a $4.20
b $8.50
c $3.60
d $9.00
e $7.30

4 Simon is at the games arcade and asks for ten 50 cent coins. He gives the cashier $10. How will he get his change?

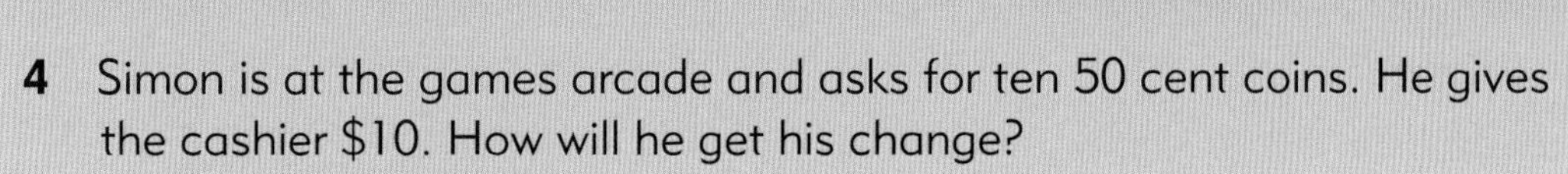

ISBN: 9780170217101

5 Sarah makes a \$2.40 purchase at the dairy and gives the assistant \$10. She wants two \$1 coins in her change. How will she get it?

6 Sydney makes the following purchases: two packets of chips at \$1.80 each; one bottle coke at \$3.40; three ice-creams at \$1.20 each.

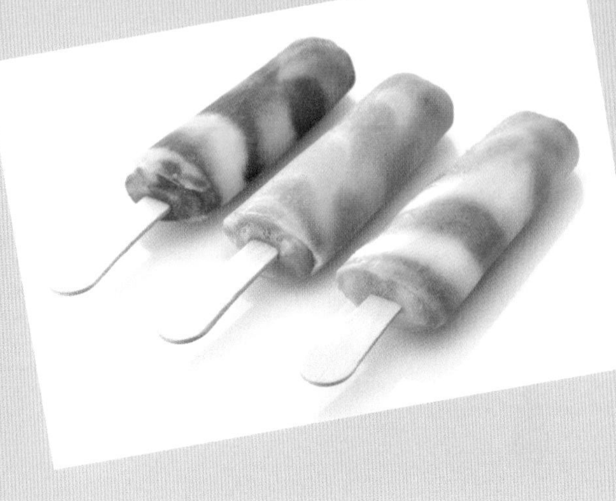

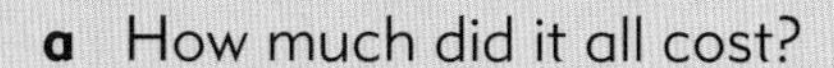

a How much did it all cost?

b If he gives the assistant \$20 note, how will he get his change?

7 Stephen bought 3 m of material at \$18.30 per m; one reel of cotton for \$3.55; one pattern at \$17.50; and 4 m of braid at 95c per m.

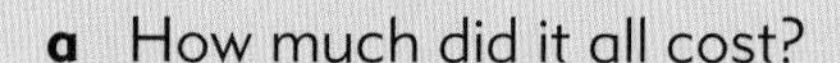

a How much did it all cost?

b If he gives the assistant a \$100 note, what would be the most convenient way of giving him change?

c How would you give Stephen his change if he gave you 1 x \$50 and 2 x \$20?

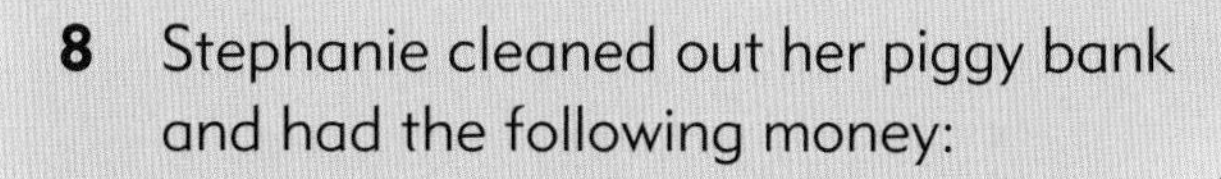

8 Stephanie cleaned out her piggy bank and had the following money:

58 x 10c
65 x 20c
1 x \$50
32 x \$1
and 49 x \$2

She cashed it at the bank and asked to be given as few notes as possible. What were they?

ISBN: 9780170217101

Pricing goods

There are several specific terms that are used when you talk about pricing goods, discounts, mark-ups, sale price, profit and loss.

Retailer: the shop owner.

Cost price: the price that the retailer buys the goods for.

Selling price or marked price: the price that the customer pays.

Sale price: the price that the customer pays in a sale after a reduction.

Discount: a reduction on the marked price.

Mark-up: the extra amount added to the cost price.

Profit: the extra amount that the retailer makes on the sale of goods.

Loss: the amount that the retailer loses on the sale of goods.

Pricing Goods

cost price	+	mark-up	=	selling price
marked price	-	discount	=	sale price
selling price	-	cost price	=	profit
cost price	-	selling price	=	loss

ISBN: 9780170217101

Exercise 15

1 Find the selling price of these goods, which have had the following percentages added to their cost price:
   - **a** Trousers $90, mark-up of 75%
   - **b** Cardigan $55.50, mark-up of 80%
   - **c** Calculator $39.90, mark-up of 100%
   - **d** Tomatoes $3.20/kg, mark-up of 12.5%
   - **e** Apples $2.50/kg, mark-up of 50%

2 What is the mark-up of these goods and what percentage is this of the cost price? (1 d.p.)
   - **a** cost price $20, selling price $30
   - **b** cost price $1.55, selling price $1.95
   - **c** cost price $48, selling price $50
   - **d** cost price 60c, selling price $1.50
   - **e** cost price $14.40, selling price $16.80

3 A shop has a sale and is offering 25% off all marked prices. What is the discount and sale price of these goods:
   - **a** Jeans marked at $119.00?
   - **b** Shirt marked at $95?
   - **c** Computer marked at $2120?
   - **d** Carpet marked at $186 per m?
   - **e** Lounge suite marked at $2560?

4 Goods in a sale are priced as below. How much is the discount and what percentage is the discount of the marked price? (1 d.p.)
   - **a** was $16.95 now $12.95
   - **b** was $67.50 now $38.00
   - **c** was $29.50 now half price

5 Find the profit or loss on these goods:
   - **a** cost price $65, selling price $75
   - **b** cost price $5, selling price $4
   - **c** cost price $80, selling price $100
   - **d** cost price $7.50, selling price $5
   - **e** cost price $108, selling price $124
   - **f** cost price 95c, selling price $1.30
   - **g** cost price $5.80, selling price $5.30

ISBN: 9780170217101

h cost price \$66, selling price \$66

i cost price \$1.20, selling price 90c

j cost price \$8.40, selling price \$8.00

6 Using your answers to 5 a-j, calculate the profit or loss as a percentage of the cost price. (1 d.p.)

7 John is an apprentice builder. DIY Hardware are having a 'GST free day' so he decided to buy the following: (all prices include 15% GST)

1 extension ladder @ \$229

2 hammers @ \$69.95 each

1 electric drill @ \$349.95

1 circular saw @ \$575.95

5 builder's rulers @ \$28.50 each

5 builder's pencils @ \$12.95 for 2

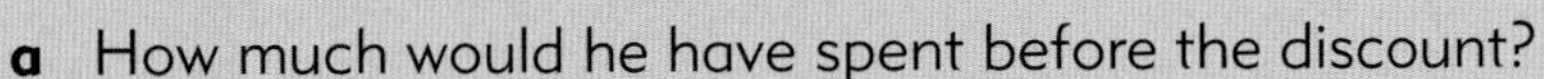

a How much would he have spent before the discount?

b How much did he save?

8 Susan decided to replant parts of her garden. Both garden centres have the same starting prices. 'Plant House' are offering 15% off everything bought and 'Qik Grow' are offering 5% off anything bought up to \$100 and 20% off the remaining amount over \$100. She wanted to buy the following:

3 bags of compost @ \$9.95 each

5 L of fertiliser @ \$19.95

8 punnets of plants @ \$2.97 each

5 rose bushes @ \$19.95 each

4 pkts of seeds @ \$3.95 each

15 m wired gardening edge @ \$23.32/4 m

a How much did she need to spend?

b Which place offered the best deal for Susan and by how much?

ISBN: 9780170217101

Putting it into practice

Achievement Standard 91026 : Apply numeric reasoning in solving problems.

Internal Assessment • 3 credits

NA5.3 Understand operations of fractions, decimals, percentages, and integers
NA5.4 Use rates and ratios
NA5.5 Know commonly used fraction, decimal and percentage conversions
NA5.6 Know and apply standard form, significant figures, rounding and decimal place value
NA6.3 Apply everyday compounding rates
NA6.4 Find optimal solutions, using numerical approaches

Student instructions: In New Zealand, Goods and Services Tax (GST) is 15%.

Useful formulae: $\text{simple interest} = \frac{PTR}{100}$

Exercise 16

1 Oliver wants to buy a TV priced at $1990. He is offered a 10% discount if he pays cash. How much will the TV cost him?

2 Jane is offered a 15% discount on a lounge suite priced at $2550. How much is her discount and what price did she pay for it?

3 A computer is normally priced at $3050 but for this week there is a 35% discount. What price would you pay if you bought one this week?

ISBN: 9780170217101

In New Zealand Goods and Services Tax (GST) is 15%. Use this to answer the following questions.

4. A car is priced at $5500 plus GST. How much is the GST and what is the selling price?

5. If the GST is to be added to the price of a car tyre, priced at $75, what is the selling price of the tyre?

6. A manufacturer sells tables and chairs for $2100 plus GST. What is the selling price?

7. A pair of jeans sells for $150 including GST. How much of the $150 is GST?

8. The average cost of my weekly grocery bill is $185. How much of this is GST?

9. Warner sells his house for $490 000 including GST. How much of this price is GST?

10. Huia invests $1000 for 3 years at 5.5% pa simple interest.
    - **a** How much interest will she earn over the three years?
    - **b** How much money does she now have?

11. Michael invests his $2500 for 4 years at 4.25% pa simple interest.
    - **a** How much interest will he earn over the four years?
    - **b** How much money does he now have?

12. Michelle invests her $5000 for three years and earns $600 in interest.
    - **a** What was the interest rate?
    - **b** How much money does she now have?

13. Ashley wants to increase her $4000 investment to $5000 over the next five years. What interest rate does she need to invest the $4000?

14. Susan had $1500 invested at 5% and the investment is now $2250. How long did she have it invested?

15. I have $1000 invested at 5% pa compounding annually for three years (answer to 3 sig figs).
    - **a** How much interest will I earn over the three years?
    - **b** How much money do I now have?

16. Heather invested her $5000 inheritance for five years earning compound interest of 4%, every six months (answer to nearest $).
    - **a** How much interest does she earn over the five years?
    - **b** How much money does she now have?

ISBN: 9780170217101

17 The bank is offering several options for investments. Which would be the best option if I want to get the greatest return for my $10 000 over the next five years?

a Simple interest at 5.5% pa.

b Compound interest at 5% pa annually.

c Compound interest at 1.5% pa quarterly.

Calculate each option and explain which would be the best option over the next five years, giving reasons for your choice.

18 Supermarkets are often offering 'best buys'. Decide which is the best buy for the following goods.

a 100 tea bags for $7.95,
50 tea bags for $3.95 or
20 teabags for $1.95?

b 4 toilet rolls for $2.25,
8 toilet rolls for $4.15 or
12 toilet rolls for $6.50?

c 1 tin spaghetti for $1.95,
3 tins spaghetti for $5.50 or
8 tins of spaghetti for $12.50?

19 Fertiliser is advertised in the following three options. Which is the best buy and why?

20 L for $9.95   50 L for $23.95   100 L for $45.95

20 It costs $25 per m^2 to paint an area up to 30m^2. then $23 for the next 20m^2. After that it costs $22 per m^2. What is the cost for the following areas needing paint?

a 40 m^2

b 150 m^2

c 380 m^2

21 The Pokais have the following income and monthly expenditure. Prepare a monthly budget for them. Is this a good budget? Explain your reasons and give possible suggestions for improvements.

- Total monthly income: $4950
- Weekly mortgage: $195
- Monthly phone and Internet: $195
- Monthly electricity: $205
- Weekly groceries: $275
- Weekly travel expenses: $100
- Weekly incidentals: $250

ISBN: 9780170217101

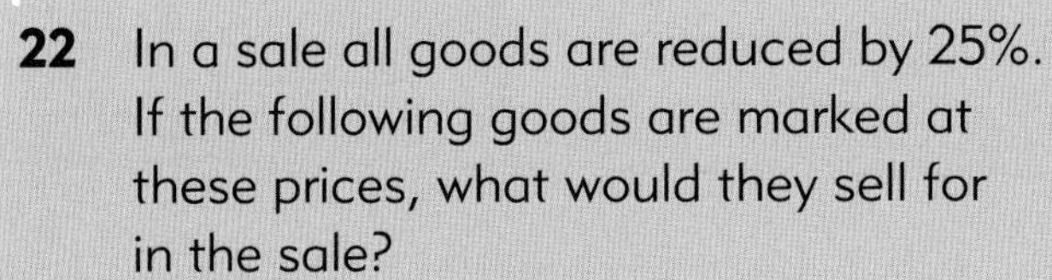

22 In a sale all goods are reduced by 25%. If the following goods are marked at these prices, what would they sell for in the sale?

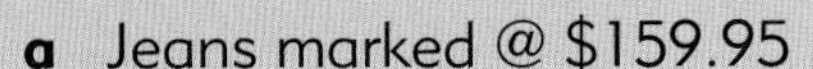

a Jeans marked @ $159.95

b Shoes marked @ $245

c Tops marked @ $99.95

23 A wholesaler buys goods for a particular price and then 75% is added to the price. What would be the selling price for these goods?

a A wholesaler buys apples for 25c per kg. What is the selling price?

b A wholesaler buys cooking oil for $8.95 per 10 L and repackages the oil into 1 L containers. How much will 1 L sell for?

c A wholesaler buys eggs for $2.50 per dozen. What is the selling price of a carton containing ten eggs?

24 Calculate: **i** the profit and/or loss of the following goods

ii the profit and/or loss as a percentage of the cost price.

a Goods costing $300 and selling for $350.

b Goods costing $250 and selling for $200.

c Goods costing $195 and selling for $395.

ISBN: 9780170217101

Achievement Standard 23739: Use number strategies to solve number problems involving decimals, percentages and fractions.

Internal Assessment • 2 credits

Assessment Conditions:
Calculators are not allowed for this closed book assessment.

Student instructions:
Solve each problem and describe or show how you solved the problem. **You must describe or show at least three different strategies for numbers 1 - 6.**

Exercise 17

1 The balance of my bank account was $1256.00. I spent $399.00 at the Farmers. What is the balance of my account?

2 The balance of my bank account was $997.55. My weekly wages of $567.50 were deposited. What is the balance of my account?

3 My annual salary is $42 159.00. How much is my weekly income?

4 My gross weekly income is $550 and 20% of this is paid as income tax. How much is the income tax and what is my net weekly wage?

5 My gross monthly income is $2040 and a ¼ of this is paid in tax. How much is my monthly net income?

6 Jane buys a TV for $1200. She has to pay a 25% deposit and 12 monthly instalments of $87.50. How much extra (interest) did she pay?

7 If I have $5000 to invest, how much interest will I earn if this money is invested for 5 years at 5.5% pa simple interest?

8 If I decide to invest my $5000 in a bank paying 4.5% pa compounding annually, how much interest will I earn after 5 years?

9 Which is the best buy for toilet paper: 12 rolls for $5.95, 10 rolls for $4.95 or 4 rolls for $1.95?

ISBN: 9780170217101

10 The following deals are advertised. Decide which is the best option when buying a TV priced at $1995 – justify your answer.

- Option A – pay cash and get a 10% discount.
- Option B – pay a 10% deposit and the balance of $100 each month for the next 20 months.
- Option C – pay no deposit but 12 monthly instalments of $265.

11 All items in the store are on sale and items have been given one of three dots. Red dots are 50% off, yellow dots are 30% off and white dots are 25% off. What are the sale prices of the following goods?

a Red dot jeans marked at $129.95

b Yellow dot tops marked at $69.95

c White dot shoes marked at $110.00

12 What is the percentage profit or loss of the cost price of the following goods?

a A computer priced at $1095, sale price $895

b A car priced at $7500, sale price $6500

c Apples bought for $0.95/kg and sold for $1.25/kg

d Cup cakes costing 25c each to make and sold for $1.20 each.

ISBN: 9780170217101

Enrichment

- You or your class may like to visit a local bank or banks and see what services they have to offer.
- The Public Relations Officer of your local bank could be invited to visit your class and talk about what the bank offers.
- You may like to do a research project on the variation in prices between your town and another town or city. Choose a variety of goods and/or services. You could investigate the reasons for the variation in prices.
- You may like to pretend you have $5000 to invest for one year. Find out all the possible places you could invest this money and how much it will earn in that time.

Calculator fun

Turn the calculator upside down to read the answer to the following clues.

1 Russian Ballet Company — 522 854 ÷ 0.5

2 A good viewing place — 90 473 142 ÷ 3

3 A good shake will make it — 210 042 ÷ 6

4 She does it on the sea shore — 23 x 15, 1283 x 45, 67 x 5, 115 469 x 5

5 Feathered friends — 17 668 x 2

ISBN: 9780170217101

Crossword

Complete the crossword below by using the following calculations.

Across

1 Sixty-eight thousand, three hundred and forty-seven

4 ¼ of 100

5 0.83 = _____ hundredths

6 Decrease 110 by 12

7 DVI

8 1000 minus 99

10 800 000 + 9 + 400 + 60

12 $3^2 \times 3^2$

13 Six more than 2 ½ dozen

14 10 x 62

16 108 + 99

17 4 m 50 cm = _____ cm

18 Eight centuries = _____ years

19 0.75 L = _____ ml

Down

2 88 hundred =

3 Sixty more than three thousand three hundred

4 10 x 4 x 7

6 1 less than 1000

7 Double 29

9 MDL + XVII

11 20 x 24

13 30 x 100

14 5 x 130

15 0.2 kg = _____ g

17 2 x 4 x 6

ISBN: 9780170217101

Word search

Find the following 20 terms in the grid below.

W	I	T	H	D	R	A	W	A	L	E	B	U	T	T	V	W	I
O	V	B	G	P	R	I	P	U	K	R	A	M	R	Q	P	J	C
P	J	P	A	Y	E	E	B	M	N	Y	X	O	W	S	P	V	H
Q	D	C	I	N	O	E	T	E	S	I	G	N	A	T	U	R	E
R	F	P	R	M	K	I	N	V	O	I	C	E	P	V	R	W	Q
T	R	A	N	S	F	E	R	B	V	E	S	Y	S	R	C	I	U
E	C	D	Q	O	K	Q	O	D	E	K	N	O	P	D	H	R	E
G	X	P	R	U	E	F	M	U	R	T	N	E	M	Y	A	P	P
D	E	P	O	S	I	T	V	Y	D	A	I	G	T	T	S	H	J
U	R	W	I	B	L	T	R	S	R	X	W	L	N	L	E	M	I
B	Q	O	T	N	U	O	C	C	A	W	F	E	E	H	J	P	N
K	N	V	S	T	C	Y	S	Y	F	K	R	C	R	E	D	I	T
U	A	C	C	O	U	N	T	S	T	S	T	E	M	E	N	T	E
L	M	W	D	H	A	L	L	B	A	N	K	C	A	R	D	U	R
T	Y	P	R	E	C	O	N	C	I	L	I	A	T	I	O	N	E
Z	S	Z	T	V	Q	A	A	H	V	Q	U	S	M	Z	P	T	S
L	U	I	O	E	C	N	A	L	A	B	O	H	K	O	S	Z	T

Bank
Money
Deposit
Signature
Purchase
Buy
Mark-up
Profit
Loss
Budget

Rent
Bankcard
Transfer
Interest
Balance
Invoice
Acc
Cash
Credit
Cheque

Butt
Statement
Loan
Payment
Withdrawal
Payee
Reconciliation
Drawer
Overdraft Account

ISBN: 9780170217101

Problem solving

1 Insert two multiplication signs and one minus sign in the following statement to make it true. Don't rearrange the numbers.

1 2 3 4 5 6 7 = 100

2 Add five matches to the six shown to make nine.

3 What are the next letters in this sequence?

BAD CEF DIG FOH

ISBN: 9780170217101

Answers

Exercise 1

1 Extra money paid for the use of money loaned.
2 $1.00
3 18 years
4 Electronic Funds Transfer at Point of Sale
5 Convenient, safe, low interest, possibly no fees, timesaving, record of transactions.
6 A loan arranged between you and your bank which can be used on your cheque account should the need arise.
7 A detailed outline of your transactions over a set period of time.
8 **a** deposit slip **b** drawer
c automatic payment **d** cirrus
e credit card **f** balance
g investment account **h** debit card

Exercise 2

1

Customer receipt

Date: 01 07 12
Total amount: 895 | 35
389813
NZSB

Deposit slip

Date: 01 07 12
NZSB
Account holder (please print): *Tom Smith* 389813
Name of depositor (please print): *Tom Smith*
Contact phone number of depositor:
Reference (limit 12 characters):
NZSB EFTPOS card number:
OR Account number: 01 0413 0834567 300 50

Notes $	250 | 00
Coins $	
Subtotal $	
Cheques (as per reverse) $	645 | 35
Total $	895 | 35

II' 50

Cheque details NZSB

Drawer	Bank	Branch	Amount
Comid Holdings	*BNZ*	*Kawerau*	645 | 35
		Total	$

NZSB New Zealand Savings Bank Limited

ISBN: 9780170217101

2

New Zealand Savings Bank NZSB

Account Payee

Pay Coastal Electronics

Date 25 May 2012

One thousand and nine hundred and ninety dollars only

$1990.00

R. Hoodwink

Cheque No Branch Sort Code Account No. Transaction Code

729808 54 8045 44586512 01

3

Customer receipt

Date 01 07 12

Total amount 169 | 50

389813

NZSB

Deposit slip

Date 01 07 12

Account holder (please print) F Parent

389813

Name of depositor (please print) L Parent

Contact phone number of depositor

Reference (limit 12 characters)

NZSB EFTPOS card number

OR Account number 01 4341 0089866 000 50

Notes $	75 | 00
Coins $	
Subtotal $	
Cheques (as per reverse) $	94 | 50
Total $	169 | 50

II' 50

Cheque details NZSB

Drawer	Bank	Branch	Amount
A. B. Smith	BNZ	Katikati	94 | 50
		Total $	94 | 50

NZSB New Zealand Savings Bank Limited

Exercise 3

1. 21 Feb 2011
2. $110.42
3. Automatic payment to a credit card
4. $1.50
5. $1874.42
6. The customer used an ATM machine owned by a different bank.
7. $2796.91
8. An automatic payment of $950 was paid to AK and DG Mandeno for Rent

Exercise 4

1. $11.21, $22.42, $44.84
2. **a** $5168.80 **b** $5168.80 **c** $10 337.60
3. **a** $1000 **b** $52.00 **c** $52.00 **d** $5408.00 **e** $54080.00

ISBN: 9780170217101

Exercise 5

1

Date	Withdrawal	Deposit	Balance
18-7-12		200.00	424.55
20-7-12	120.00		**304.55**
21-7-12		150.00	**454.55**
25-7-12	84.00		**370.55**
26-7-12	37.95		**332.60**
29-7-12	24.75		**307.85**
30-7-12		350.00	**657.85**
30-7-12	112.15		**545.70**
31-7-12		200.00	**745.70**
01-8-12	192.45		**553.25**

2 **a** NZSB Access Account
b Deborah Lee Marsh
c $17 075.49 (amount in account from previous page)
d 14 Oct 2010
e $23 329.48
f EFTPOS transaction of $55, sutomatic payment of $300 on a credit card, a NON-NZSB-ATM transaction
g $12 958.54
h $23 991.83

Exercise 6

1 $74.85, $424.15
2 9.4%
3 **a** $540 **b** $237.60 **c** Ask your teacher

Exercise 7

1

	Item (with cash price)	Deposit	Instalments	Total price	Interest (extra paid)
a	Bike, $995	15%	18 instalments @ $62	$1116 + 149.25 = **$1265.25**	**$270.25**
b	iPod, $495	20%	10 instalments @ $58.50	$585.00 + 99 = **$684**	**$189**
c	Computer, $1995	10%	24 instalments @ $85.25	$2046 + 199.50 = **$2245.50**	**$250.50**
d	Mini (car), $14 995	10%	36 instalments @ $395.00	$14 220 + 1499.50 = **$15 719.50**	**$724.50**
e	P class yacht, $2560	25%	24 instalments @ $99.95	$2398.80 + 640 = **$3038.80**	**$478.80**

2 **a** $650 **b** $2280 **c** $1155 **d** $131.25 e $44.06
3 **a** P = 100 I / TR **b** T = 100 I / PR **c** R = 100 I / PT
4 **a** 3.125% **b** 4% **c** $800 **d** $2000 **e** six months **f** four years three months
5 **a** $13 248.00 **b** $3248.00 **c** 10.8% per year
6 **a** $10 098.00 **b** $11 098.00 **c** $3098.00 **d** 38.7%
7 **a** $262.50 **b** $1042.20 **c** $1304.70
8 **a** $8950.00 **b** $2450.00 + 1022.50 = $3472.50

Exercise 8

1 $526.49 using formual $\frac{6000 \times 0.25 \times 8.5}{100}$ four successive times = $6526.49
2 **a** $1150.00 **b** $1019.86 **c** Option ii has less interest
3 **a** $2382.00 **b** $6850.00 **c** $14 775 **d** $23 080
4 **a** $499.20 **b** $2422.97 **c** $4167.68 **d** $3972.54
5 **a** $1350 **b** $1099.49 **c** $1275

Exercise 9

1 19.95% pa
2 22.2% pa
3 Getting cash from your credit card
4 0.7587 ($NZ1 = $AUS0.7587)
5 $2.64 *There were no purchases on 15 Dec but 2 processed transactions.
6 Three days
7 $21.53

ISBN: 9780170217101

Exercise 10

3 **a** $1374.43 **b** $28.00 **c** $21.53 **d** 04/02/2011 **e** $80.00 **f** No **g** $697.39
4 $200

Exercise 11

1 **a** weekly wages, Lotto win, babysitting payment, tax rebate
b rent, car insurance, phone account
c doctor's account, repairs to car, video hire, new dress, birthday present, petrol account, food, credit card account
2 **a** $160 **b** $36 **c** personal purchases, entertainment, petrol, mobile phone
3 **a** $162.50 **b** $332.50 **c** phone, clothes, hire purchase **d** $1040 **e** $1362.40
4 **a** $2352.50 **b** savings, gifts, vehicle repairs, education **c** $247.50 is left over
5 **a** $2321 **b** $479 **c** Ask your teacher
6–10 Ask your teacher

Exercise 12

1 **a** $2.63 **b** $3.21 **c** $2.65 **d** $2.17
2 Ask your teacher
3 Ask your teacher
4 **a** 338.9% **b** 697.5% **c** 1360%
5 **a** $528.26 **b** $1353.91 **c** $1369.57 **d** $65.22
6 Income $26 360 per month, expenses $11 769.60 per month
7-11 Ask your teacher

Exercise 13

2 **a** 1 kg @ $27.20 **b** 830 g @ $2.05 **c** 1 kg @ $9.95 **d** 425 g @ $1.40
e One pair @ $2.99 **f** 375 ml @ $1.40 g pkt 200 @ $2.05
4 20 kg @ $6.95 is cheaper
5 Saving of $5.97
6 The new tyre
7-8 Ask your teacher

Exercise 14

1 2 x 50c, 1 x 50c + 2 x 20c + 1 x 10c, 1 x50c, 1 x 20c + 3 x 10c, 1 x 50c + 5 x 10c, 5 x 20c, 4 x 20c + 2 x 10c, 3 x 20c + 4 x 10c, 2 x 20c + 6 x 10c, 1 x 20c + 8 x 10c, 1 x 20c + 8 x 10c, 10 x 10c.
2 4 x $5, 2 x $10, 1 x $20, 2 x $5 and 1 x $10
3 **a** 4 x 20c and $5 **b** 1 x 50c and $1 **c** 2 x 20c, $1 and $5 **d** $1 **e** 1 x 20c, 1 x 50c and $2
4 10 x 50c and $5
5 1 x 10c, 1 x 50c, 2 x $1 and $5
6 **a** $10.60 **b** 2 x 20c, 2 x $2 and $5
7 **a** $79.75 ($79.80) **b** 1 x 20c, 1 x $20 **c** 1 x 20c, 1 x $10 and give back 3 x $50
8 1 x $100, 1 x $50, 2 x $20, 1 x $5, 1 x $2, 4 x 20c, 1 x $1

Exercise 15

1 **a** $157.50 **b** $99.90 **c** $79.80 **d** $3.60/kg **e** $3.75/kg
2 **a** $10, 50% **b** 40c, 25.8% **c** $2, 4.2% **d** 90c, 150% **e** $2.40, 16.7%
3 **a** $29.75, $89.25
b $23.75, $71.25
c $530, $1590
d $46.50, $139.50
e $640, $1920
4 **a** $4, 23.6%
b $29.50, 43.7%
c $14.75, 50%
5 **a** $10 (P) **b** $1 (L) **c** $20 (P) **d** $2.50 (L) **e** $16 (P)
f 35c (P) **g** 50c (L) **h** 0 (-) **i** 30c (L) **j** 40c (L)
6 **a** 15.4% **b** 20% **c** 25% **d** 33.3% **e** 14.8%
f 36.8% **g** 8.6% **h** 0% **i** 25% **j** 4.8%
7 **a** $1469.68 **b** $191.70
8 **a** $276.56 **b** Qik Grow by $1.17

Putting it all together
Exercise 16
AS 91026

1 $1791
2 $382.50, $2167.50
3 $1067.50 x $1982.50
4 $825, $6325
5 $86.25
6 $2415
7 $127.50 x $19.57
8 $24.13

ISBN: 9780170217101

9 $63 913.04
10 **a** $165 **b** $1165
11 **a** $425 **b** $2925
12 **a** 4% **b** $5600
13 5%
14 10 years
15 **a** $157.63 **b** $1157.63
16 **a** $2401.22 **b** $7401.22
17 Option C offers the best interest but discuss with the teacher the other options.
18 **a** 50 bags $3.95 **b** 8 toilet rolls $4.15 **c** 8 tins $12.50
19 100 L $45.95
20 **a** $980 **b** $3410 **c** $8470
21 Income $4950, outgoings $3680, surplus $1270 per month. Discuss with your teacher.
22 **a** $119.96 **b** $183.75 **c** $74.96
23 **a** 44c **b** $1.57/L **c** $3.65
24 **a i** P= $50 **ii** 16.7%
b i L= $50 **ii** 20%
c i P= $200 **ii** 102.6

Excercise 17
AS 23739

1 (1256 – 399) or, (1256 – 400 +1) or, (1200 + 400 + 55)
2 (997.55 + 567.50) or (997.55 + 500 + 60 + 7 + 50c) or (1000 + 567.50 – 2.45)
3 (42 160 ÷ 52) or (42 160 ÷ 26 ÷ 2) or (40 000 ÷ 52, 2 000 ÷ 52, 100 ÷ 52, 60 ÷ 52)
4 (20% of 550) or (10% of 550 x 2) or $110 tax, net wage $440 or 550 ÷ 5
5 (1/4 of 2040) or (0.25 x 2040) or (1/2 of 2040 then ½ of this) Monthly income $1530
6 (25% of 1200) or (1200/4) or (10% of 1200, then x2 plus a half of 10%) Interest $150 + 87.5 x 12
7 $1375
8 $1230.90
9 4 rolls for $1.95
10 Option A is the cheapest
11 **a** $64.98 **b** $48.97 **c** $82.50
12 **a** L - 18.3% **b** L - 13.3% **c** P - 31.6% **d** P – 380%

Enrichment
Calculator fun

1 Bolshoi
2 Hillside
3 Loose
4 She sells see shells
5 Geese

Crossword

1 6	2 8	3 3	4	7	■	4 2	5
■	5 8	3	■	■	6 9	8	■
7 5	0	6	■	■	8 9	0	9 1
10 8	0	0	11 4	6	9	■	5
■	■	■	12 8	1	■	13 3	6
■	14 6	15 2	0	■	16 2	0	14 7
17 4	5	0	■	■	■	0	■
18 8	0	0	■	19 7	5	0	■

ISBN: 9780170217101

Word search

W	I	T	H	D	R	A	W	A	L	E	B	U	T	T	V	W	I
O	V	B	G	P	R	I	P	U	K	R	A	M	R	Q	P	J	C
P	J	P	A	Y	E	E	B	M	N	Y	X	O	W	S	P	V	H
Q	D	C	I	N	O	E	T	E	S	I	G	N	A	T	U	R	E
R	F	P	R	M	K	I	N	V	O	I	C	E	P	V	R	W	Q
T	R	A	N	S	F	E	R	B	V	E	S	Y	S	R	C	I	U
E	C	D	Q	O	K	Q	O	D	E	K	N	O	P	D	H	R	E
G	X	P	R	U	E	F	M	U	R	T	N	E	M	Y	A	P	P
D	E	P	O	S	I	T	V	Y	D	A	I	G	T	T	S	H	J
U	R	W	I	B	L	T	R	S	R	X	W	L	N	L	E	M	I
B	Q	O	T	N	U	O	C	C	A	W	F	E	E	H	J	P	N
K	N	V	S	T	C	Y	S	Y	F	K	R	C	R	E	D	I	T
U	A	C	C	O	U	N	T	S	T	S	T	E	M	E	N	T	E
L	M	W	D	H	A	L	L	B	A	N	K	C	A	R	D	U	R
T	Y	P	R	E	C	O	N	C	I	L	I	A	T	I	O	N	E
Z	S	Z	T	V	Q	A	A	H	V	Q	U	S	M	Z	P	T	S
L	U	I	O	E	C	N	A	L	A	B	O	H	K	O	S	Z	T

Problem solving

1 123 x 4 – 56 x 7 = 100

2

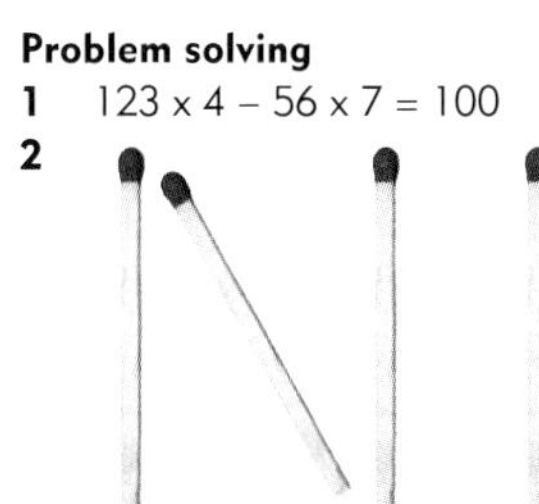

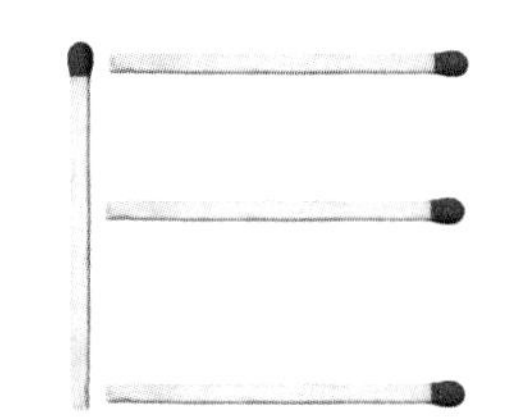

3 CUJ

ISBN: 9780170217101